WASTEWATER TREATMENT IN THE FISHERY INDUSTRY

WASTEWATER TREATMENT IN THE FISHERY INDUSTRY

WASTEWATER TREATMENT IN THE FISHERY INDUSTRY

J F Gonzalez

Reprinted by arrangement with the
Food and Agriculture Organization of the United Nations

by

Daya Publishing House

A Division of

Astral International Pvt Ltd

New Delhi - 110 002

"This book was originally published by the Food and Agriculture Organization of the United Nations (FAO) as Vaccine Manual, FAO Animal Production and Health Series No. 35".

"The designations employed and the presentation of material in this publication do not imply the expression of any opinion, whatsoever on the part of the Food and Agriculture Organization of the United Nations concerning the legal status of any country, territory, city or area or of its authorities, or concerning the delimitations of its frontiers or boundaries."

'The designations 'developed' and 'developing' economies are intended for statistical convenience and do not necessarily express a judgment about the stage reached by a particular country, territory or area in the development process.'

First Indian Edition, 2005
Reprinted, 2024
ISBN: 9788170354055

Published by : **Daya Publishing House**
A Division of
Astral International Pvt Ltd
4736/23, Ansari Road
Darya Ganj, New Delhi - 110002
Phone: 011-43549197
Email: info@astralint.com
Web: www.astralint.com

Printed at : **Chawla Offset Printers,**
Delhi-110 052

PREPARATION OF THIS DOCUMENT

Concern on environmental issues is growing in both developed and developing countries. In particular, fishery industries (capture and aquaculture) suffer from pollution produced by other sectors of the society (industry, agriculture and urban sewage). As a consequence, fishery industries are becoming increasingly aware of their role as a potential source of pollution; and they must show a responsible attitude regarding waste from their fishery industry if they expect other sectors to reduce their pollution of waterbodies. Further, a number of Governments have decided to pass rules enforcing the fishery industries to treat their effluents in an appropriate way (e.g., Thailand and Venezuela). Fish technologists, particularly in developing countries, must become informed on this issue and be in position to advise Governments and the fishery industry on fishery wastewater treatment. This need, identified in several FAO/DANIDA Workshops on Fish Technology carried out by project GCP/INT/609/DEN (formerly GCP/INT/391/DEN), motivated the preparation of this document.

In addition to the classic approach of utilizing solid fish wastes in the production of fish meal and oil, the current tendency of the fishery industry regarding wastewater moves in two complementary directions. On the one hand, the aim is to reduce wastes as much as possible by better fish handling and processing and by a rational use of water; this approach is usually called "clean production". On the other hand, the objective is to search for wastewater treatment processes that could offset the costs involved, totally or in part, through the recovery of substances, in particular protein and fat, from wastewater. The two approaches are complementary in that "clean production" reduces the volume of effluents and makes them more concentrated, thus improving the overall economics of the recovery process. This paper deals specifically with wastewater characterization (explained in section 2), treatment and recovery.

The present document was originally prepared for use at courses and workshops on fish technology, in which the participants have a basic knowledge of chemistry and microbiology. However, for people with practical experience working in the fishery industry, the document provides the necessary background information on fisheries wastewater treatment. FAO has decided to publish it as an FAO Fisheries Technical Paper rather than as a project publication, to allow for extensive diffusion in view of the worldwide relevance of the subject involved. The preparation, editing and printing of this paper have been financed by the FAO/DANIDA Training Project on Fish Technology and Quality Assurance (GCP/INT/609/ DEN). The outline was proposed and the technical editing were done by Mr H.M. Lupin (Project Manager GCP /INT/609/DEN).

Distribution

FAO Fisheries Department.
FAO Regional and Sub-Regional Fisheries Officers
Selector Handling and Processing
FAO Fisheries field projects
FAO Representative Offices
DANIDA

González, J.F.
Wastewater treatment in the fishery industry.
FAO Fisheries Technical Paper. No. 355. Rome, FAO. 1995. 52p.

ABSTRACT

This document presents fish processors with an introduction to the concepts of fisheries wastewater characterization and the various types of treatment utilised. The topics dealt with include characterization analysis, primary treatment, biological treatment, physicochemical treatment including disinfection, sludge treatment and disposal. Economic considerations are among the most important parameters that influence the final decision on which process should be chosen for wastewater treatment and these are presented in the final section.

ACKNOWLEDGEMENTS

The author wishes to acknowledge the assistance and resources provided by the National University of Mar del Plata (UNMdP) and the National Research Council of Argentina (CONICET).

TABLE OF CONTENTS

LIST OF TABLES

LIST OF FIGURES

1. INTRODUCTION

This document provides an introduction and orientation to fish processors on the concepts of fisheries wastewater characterization and to the various types of treatment that can be applied.

Similar to most processing industries, fish processing operations produce wastewater, which contains active organic contaminating organisms in soluble, colloidal and particulate form. Depending on the particular operation, the degree of contamination may be small (e.g., washing operations), mild (e.g., fish filleting), or heavy (e.g., bloodwater drained from fish storage tanks).

The magnitude of the problem created by these wastewater streams cannot be generalized, and is strictly related to the individual case, since the impact depends, *inter alia* on the strength of the effluent, the rate of discharge, and the assimilatory capacity of the receiving waterbody.

Section 2 (Characterization) explains the main pollution parameters that may have to be taken into account when characterizing a wastewater and also when evaluating the quality or efficiency of a wastewater treatment system.

Section 3 (Primary treatment) deals first with the most simple operations that can be performed on an effluent in order to reduce its contaminant load, such as screening and settling. The separation of oil and grease is explained, followed by a description of flotation.

Section 4 is devoted to the biological treatments. The concepts of aerobic and anaerobic processes are explained, followed by description of the most common operations of biological treatment. The aeration systems are explained in detail, since they are a key factor for the success of aerobic treatments.

Section 5 explains the most common operations of physicochemical treatment: first the coagulation and flocculation used essentially to remove colloidal matter present in wastewaters; and then the disinfection processes such as chlorination and ozonation.

Section 6 deals with the most simple treatment processes for sludges generated during the primary biological and physicochemical treatment.

Section 7 discusses economic issues related to wastewater treatment processes.

2. WASTEWATER CHARACTERIZATION

As in most wastewaters, the contaminants present in fisheries wastewaters are an undefined mixture of substances, mostly organic in nature. Since a detailed analysis for each component is neither useful nor practically possible, most of the analyses give an overall measure of the degree of contamination present. In the text that follows an estimation of organic content is discussed since this is one of the main pollution parameters.

2.1 Physicochemical Parameters

2.1.1 pH

The pH itself is not a contaminant but is important as a characterization parameter since it may reveal contamination or indicate the need for its correction before biological treatment of the wastewater. Effluents from fish processing plants are seldom acidic and are usually close to 7 or alkaline. This value is generally due to the decomposition of the proteinaceous matter and emission of ammonia compounds.

2.1.2 Solids content

Solids may be present either in dissolved or suspended form. Suspended solids are of primary concern since they are objectionable on several grounds: those that settle can do so in the wastewater ducts, reducing their capacity; or if they settle in the receiving waterbody they may affect the bottom-dwelling flora and the food chain; if they float, the light that enters from the surface is reduced, and those that remain suspended reduce the amount of light that enters the water thereby affecting wildlife. The solids which settle are usually measured with an Imhoff cone (see Figure 2.1), in which a known amount of water (e.g., 1 litre) is placed. The solids which settle are estimated at fixed times, usually after 10 minutes and after 2 hours. The admissible amounts that can be discharged depend on each regulation, but discharge of wastewaters is usually not permitted if they contain solids which settle after 10 minutes.

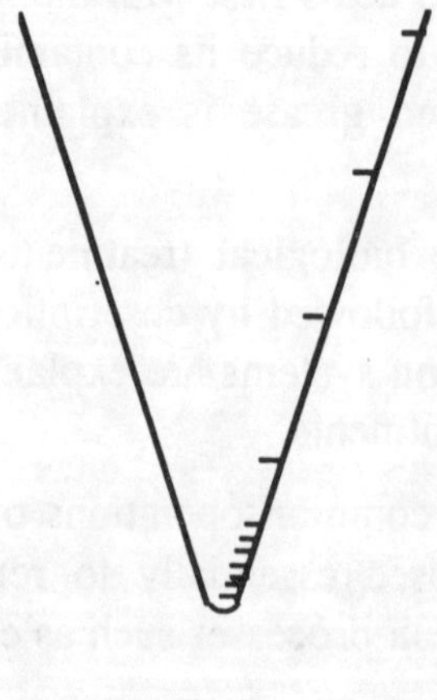

Figure 2.1. Imhoff cone used to measure settling solids

The suspended solids are measured by passing a well-mixed sample through a fibreglass filter. The weight of suspended solids can be calculated by weight differences between the filter alone and the weight plus retained solids after drying.

Soluble solids can be measured by gravimetry after evaporation of the filtrate of a sample of known volume, but are generally not checked even though in effluents with a low degree of contamination they can be significant. They depend not only on the degree of contamination but also on the quality of the supply water: in one analysis of fish filleting wastewater it was found that 65% of the total solids present in the effluent were already in the supply water (González *et al.*, 1983).

2.1.3 Temperature

With the exception of wastewaters from cooking and sterilization processes in a canning factory, fisheries wastewaters do not discharge wastes above ambient temperatures. The temperature of the receiving waterbody must not increase by more than 2°C or 3°C, since greater increases in temperature may affect the population balance and also reduce the solubility of oxygen, thereby threatening the survival of some forms of aquatic wildlife. Wastewaters from canning operations should be cooled if the receiving waterbody is not large enough to restrict the change in temperature to 3°C.

2.1.4 Odour

Odours in fisheries wastewaters are caused by the decomposition of the organic matter that emits volatile amines, diamines and sometimes ammonia.

In wastewater that has become septic, the characteristic odour of hydrogen sulphide may also develop. Odours are very important in relation to the public perception and acceptance of any waste treatment plant. Although relatively harmless, they may affect people by inducing stress and nausea.

For measurement or estimation of odour, a test panel is exposed to odours diluted with clean air. The number of dilutions needed to reduce the odour intensity to its detectable limit is called as the detectable threshold odour concentration. The technique for determination of the threshold odour can be found in the Standard Methods (1989), but as it is complicated and subject to errors due to adaptation of the test subjects, subjectivity and sample modification, it is seldom used.

2.2 Organic Content

The organic content of the wastewater can be estimated in several ways. The most common are the oxygen demand methods, although organic carbon measurement may also be used. The first estimate is the amount of oxygen that will be needed to stabilize the organic content of the effluent. The two most common methods are the biochemical oxygen demand and the chemical oxygen demand, which are discussed in detail in the following paragraphs.

2.2.1 Biochemical oxygen demand

The biochemical oxygen demand, also known as BOD, estimates the degree of contamination by measuring the oxygen required for the oxidation of organic matter by the aerobic metabolism of the microbial flora. In fisheries wastewaters, this oxygen demand originates mainly from two sources: the carbonaceous compounds which are used as substrate by the aerobic micro-organisms, and the nitrogen-containing compounds which are normally present in fisheries wastewaters, such as proteins, peptides and volatile amines. The most common procedures are described below.

2.2.1.1 Dilution method

This is the most common procedure. It basically consists of diluting the wastewater (in a ratio that depends on its strength) with a nutrient (mineral salts) solution saturated with air, storing the dilution in airtight bottles, and measuring the dissolved oxygen at the start of the analysis and at periodic intervals. A five-day period is generally monitored and the BOD is then reported as BOD_5. Detailed procedures for its analysis are given elsewhere (Standard Methods, 1989), and a simplified scheme is given in Table 2.1.

Table 2.1. Simplified description of the BOD_5 test

BOD TEST	
1.	Obtain sample and transport immediately to laboratory
2.	Dilute samples with nutrient solution so that the maximum BOD is 6 mg /l
3.	Inoculate with adapted bacteria (see discussion in text)
4.	Fill the BOD bottles with the diluted effluent and seal them
5.	Fill blank bottles with dilution water only
6.	Determine by duplicate the dissolved oxygen (DO) in samples and in control at the start of test
7.	Incubate samples for 5 days in dark at 20°C
8.	Determine by duplicate the DO in samples and in control
9.	Calculate the net change in dissolved oxygen as: (change in DO of sample)-(change in DO of control). The BOD_5 is the net change in DO multiplied by the dilution factor used in step 2

Key points to be taken into account for a reliable BOD test will be highlighted. The first relates to the microbial population: since the BOD analysis involves biodegradation of organic matter by microbes, these must be present in the BOD bottles. In most fisheries effluents, the microbial count is such that this does not present a problem (a possible exception might be the wastewater from cooking or sterilization operations if the sample is taken either too hot or too close to the discharge from the autoclave). If the microbial count in the seed is not sufficient, the BOD may be underestimated because of the lag-time needed for the bacteria to proliferate. The same may happen if a non-adapted seed of bacteria is used. This underestimation may occur because of the extra time needed by the micro-organisms to adapt and/or grow to the extent needed to degrade the organic matter at a noticeable rate (see Figure 2.2). This is the reason why the oxygen consumption in the BOD should be followed daily, and not only at the start of the analysis but for a 5-day period. This second practice does not allow for detection of the pitfalls in the procedure mentioned above.

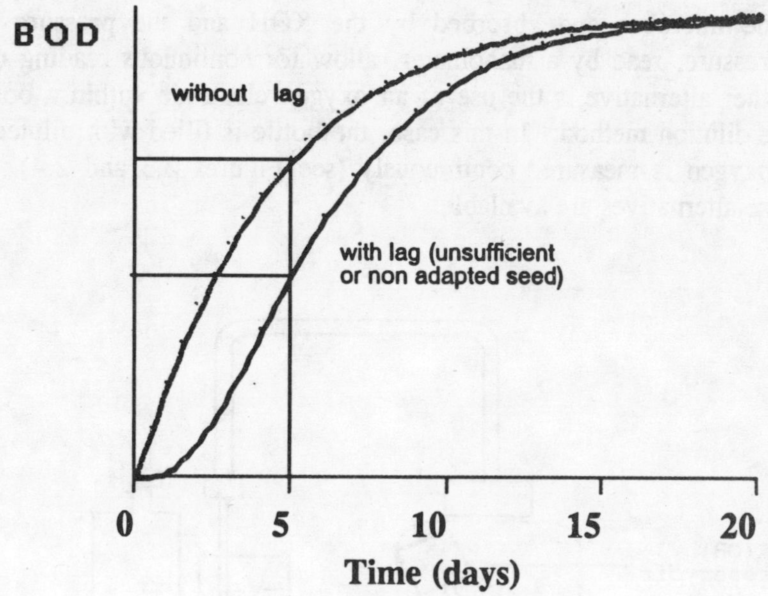

Figure 2.2. Typical BOD curves. Note that underestimation may occur if inappropriate seed is used

Another phenomenon that may occur is nitrification, since fisheries wastewaters contain appreciable amounts of proteinaceous compounds or products resulting from its degradation. This causes an additional oxygen consumption. Although nitrification begins to take place after five days, it may occur earlier depending on the amounts and form of nitrogen present and the composition of the microbial flora.

To determine the BOD which is caused solely by carbonaceous matter, chemicals (such as allyl thiourea, methylene blue or 2-chloro-6-(trichloromethyl) pyridine) may be added to inhibit nitrification in the BOD analysis (Metcalf and Eddy Inc., 1979). However, since nitrification will impose an oxygen demand on the receiving waterbody, it should be taken into account as part of the total oxygen demand of a wastewater.

The standard test calls for a 5-day incubation period at a constant temperature of 20°C in the dark. The latter is needed to avoid algae growth that might interfere if the bottles are exposed to light during incubation.

As an alternative for highly contaminated wastes, lower incubation temperatures or shorter periods may be used, but the conditions of analysis should be clearly explained together with the results.

2.2.1.2 Alternatives to the dilution method for BOD determination

Although it is standard procedure, the dilution method involves inconveniences which have prompted the search for alternate, and simpler and more rapid procedures. Among the inconveniences are: the relatively long amount of time and the glassware needed for a reliable determination, the need to prepare several dilutions (with the consequent increase in the chance of errors) and the need for an acclimated seed.

An alternative to the dilution method is the respirometric method, which uses a bottle partially filled with the wastewater to be tested. The sample is continuously stirred, and there is a reservoir with alkali (usually KOH). As the oxygen is consumed, CO_2 is liberated by the microbes and absorbed by the KOH and the pressure varies. The variations in pressure, read by a manometer, allow for continuous reading of the oxygen demand. Another alternative is the use of an oxygen electrode within a bottle similar to that used in the dilution method. In this case, the bottle is filled with diluted sample, and the dissolved oxygen is measured continuously (see Figures 2.3 and 2.4). Commercial versions of these alternatives are available.

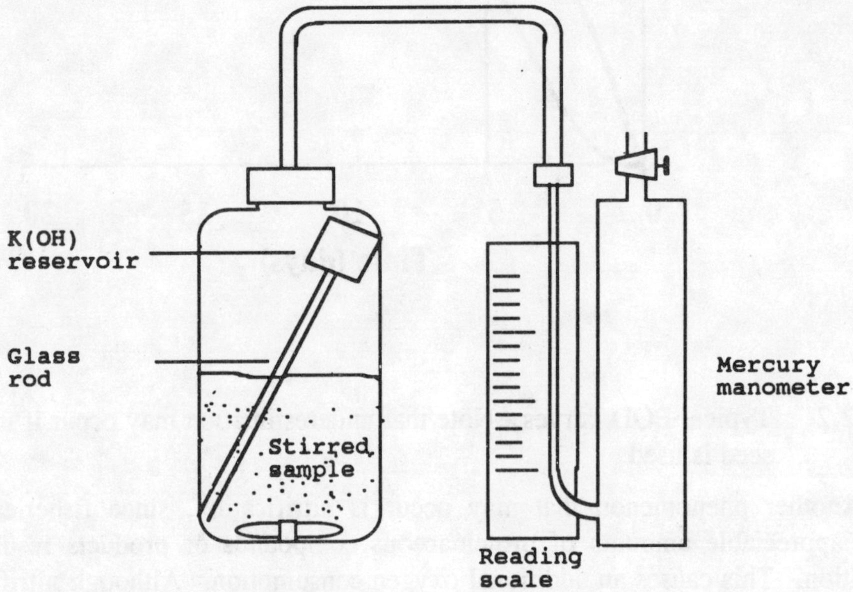

Figure 2.3. Manometric device to estimate the BOD of a wastewater

Figure 2.4. Dissolved oxygen electrode assembled into a BOD bottle

One of the major inconveniences of the BOD_5 is the time it requires to obtain results. Five days render it unsuitable for control purposes of a treatment plant or for quick monitoring since the residence time of the wastewater in the treatment facility is only of several hours. This is one of the main reasons why chemical oxidation tests (discussed below) were developed.

2.2.2 Chemical oxygen demand

Due to the inconveniences of the BOD_5, alternative methods are applied to estimate the organic content of wastewater, the most common being the chemical oxygen demand (COD). Two methods are known, both based on the chemical oxidation of the matter present: one is the oxidation by means of permanganate (sometimes known as consumed oxygen) and oxidation by the dichromate ion. Oxidation by means of permanganate was the standard method until 1965 when it was replaced by the dichromate method.

The COD analysis, by the dichromate method, is more commonly used to control and continuously monitor wastewater treatment systems. The COD of an effluent is usually higher than the BOD_5 since the number of compounds that can be chemically oxidized is greater than those that can be degraded biologically. It is also common to make a correlation of BOD_5 versus COD and then use the analysis of COD as a rapid means of estimating the BOD_5 of a wastewater. This may be convenient since only about three hours are needed for a COD determination, while a BOD_5 takes at least 5 days. However, this procedure can be used only for specific situations where there is low variability in the composition of a wastewater, and the results of a system cannot be used reliably in other cases.

The method of COD which uses dichromate as oxidant is carried out by heating under total reflux a wastewater sample of known volume in an excess of potassium dichromate ($K_2Cr_2O_7$) in presence of sulphuric Acid (H_2SO_4) for a fixed period (usually two hours) in presence of silver sulphate (Ag_2SO_4) as catalyst. The organic matter present is oxidized and, as a result, the dichromate ion (orange colour) is consumed and replaced by the chromic ion (green colour):

$$Cr_2O_7^{2-} + 14H^+ + 6e^- \rightarrow 2Cr^{3+} + 7H_2O$$

The COD is calculated by titrating the excess of dichromate or by spectrophotometrically measuring the Cr^{3+} ion at 606 nm. Another possibility is to measure the excess dichromate spectrophotometrically at 440 nm. Titration requires more work but is considered more precise.

The presence of silver sulphate as catalyst is needed for complete oxidation of aliphatic carbon compounds. The standard method implies cooling of the sample after the two-hour digestion period, adding a few drops of indicator (ferroin) solution and titrating the excess dichromate with a solution of ferrous ammonium sulphate of known concentration, until the colour changes from brilliant green to reddish brown. The titration reaction corresponds to the oxidation of the ferrous ammonium sulphate by the dichromate:

$$Cr_2O_7^{2-} + 14H^+ + 6Fe^{2+} \rightarrow 2Cr^{3+} + 6Fe^{3+} + 7H_2O$$

The change in colour corresponds to the formation of the complex ferrous ion phenantroline which occurs when all the dichromate ion has been reduced to Cr^{3+}:

$$(Fe(C_{12}H_8N_2)_3)^{3+} + e \rightarrow (Fe(C_{12}H_8N_2)_3)^{2+}$$

Ferric Phenantroline	Ferrous Phenantroline
(green-blue)	(reddish brown)

A common interference factor in the COD test is the presence of chlorides. If seawater is used at some point in the processing or salt brines are used for some "curing" operations, chlorides will most probably appear in the wastewater causing interference while they are oxidized by the dichromate:

$$Cl^- + Cr_2O_7^{2-} + 14H^+ \rightarrow 3\,Cl_2 + 2Cr^{3+} + 7H_2O$$

This interference causes erroneously high values of COD which can be prevented by the addition of mercuric sulphate ($HgSO_4$) which reacts to form mercuric chloride and precipitates:

$$Hg^{2+} + 2\,Cl^- \rightarrow HgCl_2\downarrow$$

This complexation technique has been used with success in high salinity waters (González, 1986, Bauman, 1974) and seems to be satisfactory provided the ratio Cl^- to $HgSO_4$ is at least 10 to 1.

The nitrite ion also exerts COD, but since it is present in very low amounts, its influence is generally neglected.

Detailed procedures for the reflux method of measuring the COD are given in the Standard Methods (1989).

Another method of the COD determination is that of the closed reflux method (Standard Methods, 1989; González, 1986; Jirka and Carter, 1975). In this case, a small volume of sample is heated with concentrated dichromate solution in presence of silver sulphate and mercuric sulphate. The reaction takes place in culture tubes with PTFE-lined screw caps. Heating proceeds for usually shorter times, at higher temperatures than in the open reflux method, and the COD is estimated spectrophotometrically. In this case, commercial versions of the apparatus are available as well as kits with solution ampoules and pre-measured reagents.

The open reflux method is recommended for a wide range of wastes and a large sample size can be used. The closed reflux method is more economical in terms of reactants. However, in homogenizing the samples before the analysis to obtain reproductive results, care should be taken especially in samples with suspended solids as in the case of fisheries wastewaters.

2.2.3 Other methods for estimation of organic content

2.2.3.1 Total organic carbon (TOC)

This method is based on the combustion of organic matter to carbon dioxide and water. After separation of water, the combustion gases are passed through an infra-red analyzer and the response is recorded. A diagram of this system is shown below:

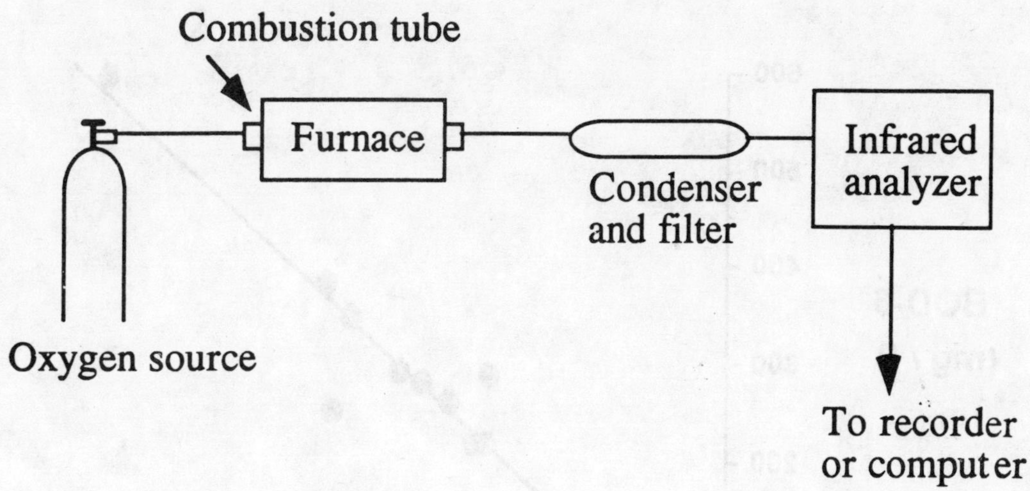

Figure 2.5. Diagram of a TOC analyzer

The signal recorded is proportional to the carbon dioxide formed which is directly proportional to the carbon concentration in the sample injected into the combustion tube. The oxidation of the sample proceeds either by heating to almost 1000°C or by intense UV radiation.

Since a wastewater contains carbonates or bicarbonates that are normally present in the supply water, these may be eliminated by acidification of the sample and purging with an inert gas prior to injection. One of the commercial versions available uses a different approach to differentiate between organic and inorganic carbon: it has two combustion tubes, one working at 150°C and the other at about 950°C. The first oxidizes the inorganic carbon, while the second oxidizes all the carbon present. The TOC is obtained from the difference between the readings. A major inconvenience for the widespread use of the TOC analysis is the relatively high cost of the apparatus.

2.2.4 Relationships between estimates of the organic content

Since all of these procedures are based on oxidation of the substances present in the wastewater, in principle, a correlation may be established between their results.

However, extreme care must be taken when developing these correlations since each of these analyses is affected by different factors. For example, the BOD_5 assay depends on many factors, such as adaptation of the seed and degree of dilution and proper pH; a minor fraction of the COD may be due to the presence of inorganic substances that reduce dichromate; there are compounds that are not degraded during the BOD_5 but are chemically oxidized such as toxic substances; there are substances resistant to oxidation by dichromate such as pyridine and acetic acid (although these are not normally present in fisheries wastewaters). Despite these factors, a correlation may be developed between two or more of the indicators of organic load (see Figure 2.6). They are not applicable to situations different from that in which it was developed and, therefore, extrapolation is not reliable and should be periodically re-checked.

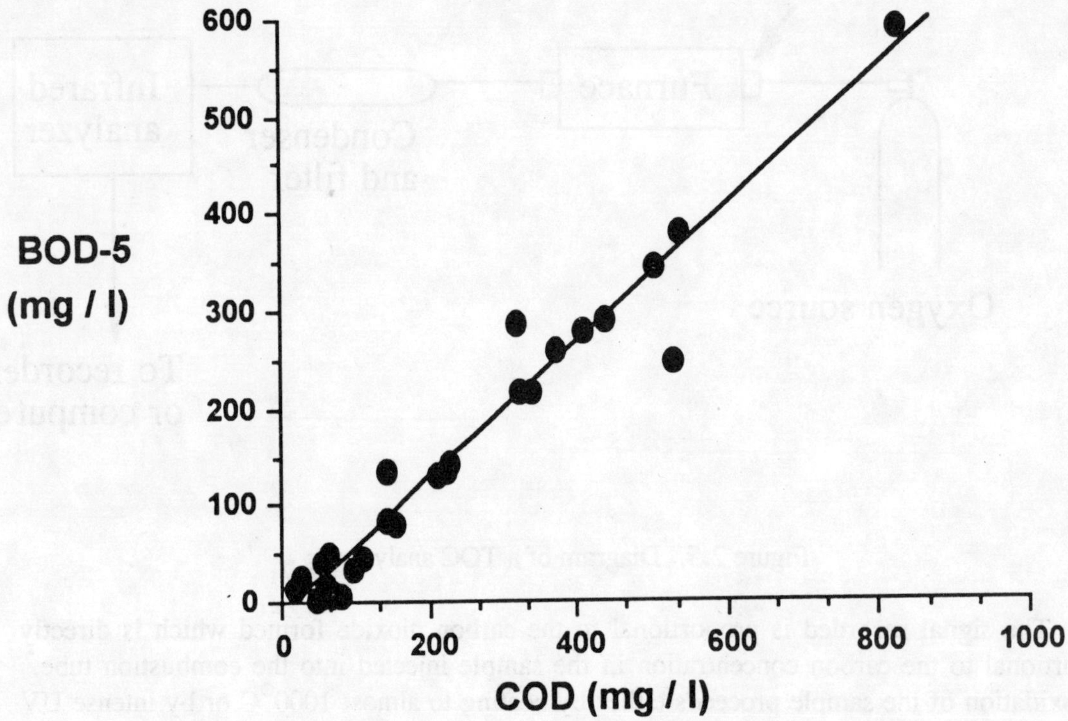

Figure 2.6. BOD_5 versus COD relationship for raw and treated fish filleting wastewater (González, 1985)

2.2.5 Oil and grease

The presence of oil and grease in an effluent is mainly due to the processing operations (e.g., canning) and, to a lesser extent, on the species being processed. In any case, they should be removed since they usually float and affect the oxygen transfer to the water and also objectionable from an aesthetic point of view. They may also cling to wastewater ducts and reduce their capacity in the long term. Generally, they are measured by extraction with solvent (Standard Methods, 1989).

2.3 Nitrogen and Phosphorous

Both nitrogen (N) and phosphorus (P) are of environmental concern because they are utrients, and if present in excess they may cause proliferation of algae (algal bloom) and affect the rest of the wildlife in a waterbody.

Although N and P are normally present in the fisheries wastewater, their concentration is minimal in most cases. For biological treatment, a range of N to P in the order of 5 to 1 is recommended for proper growth of the biomass (Metcalf and Eddy Inc., 1979; Eckenfelder, 1980).

Effluent	BOD	COD	Grease/oil	Total solids	Suspended solids	Ref. No.
Finfish processing (manual)	**3.32 kg/t**		**0.348 kg/t**		**1.42 kg/t**	**1**
Finfish processing (mechanic)	**11.9 kg/t**		**2.48 kg/t**		**8.92 kg/t**	**1**
Patagonian hake filleting	**327-1063 mg/1**	**550-1250 mg/1**	**8.3-79.9 mg/1**	**1290-4300 mg/1**		**2**
Herring filleting	**3428-10000 mg/l**		**857-6000 mg/l**			**3,4**
Tuna canning	**6.8-20 kg/t**		**1.7-13 kg/t**		**3.8-17 kg/t**	**1**
Sardine plant	**9.22 kg/t**		**1.74 kg/t**		**5.41 kg/t**	**1**
Blue crab plant	**4.8-5.5 kg/t**		**0.21-0.3 kg/t**		**0.7-0.78 kg/t**	**1**
Clam plant (mechanic)	**5.14 kg/t**		**0.145 kg/t**		**10.2 kg/t**	**1**
Clam plant (conventional)	**18.7 kg/t**		**0.461 kg/t**		**6.35 kg/t**	**1**
Fish meal plant	**2.96 kg/t**		**0.56 kg/t**		**0.92 kg/t**	**1**
Fishpumping water	**2100-7400 mg/1**		**10-1504 mg/1**	**14.5-48.2 mg/1**		**5**
Fishpumping water	**3050--67200 mg/1**		**1300--17200 mg/1**	**18.4-64.9 mg/1**		**8**
Bloodwater (fishmeal plant)	**23500-34000 mg/l**	**93000 mg/l**	**0%-1.92%**	**2.4%-6.3%**		**6,7**
Stickwater (fishmeal plant)	**13000-76000 mg/1**		**60-1560 mg/1**	**25-62 mg/1**		**5**

References : 1: Middlebrooks (1979); 2: González (1983); 3: Sorensen(1974); 4: Herborg (1974); 5: Cuadros and González (1991); 6: Parín *et al.*, 1979; 7: Civit *et al.*, 1982 8: Nemerow (1971)

2.4 Characteristics of Fish Processing Wastewaters

The degree of pollution of a wastewater will depend on several parameters of which the most important are the operation being carried out and the fish species being processed. Considering only one type of operation, the operating routine in each factory also exerts strong influence on the wastewater characteristics. Table 2.2 lists some of the ranges reported in the literature.

A particular parameter may present a wide range of numerical values. Table 2.2 emphasizes that there is no substitute for a direct determination of contamination parameters in the effluent being considered. Moreover, given the variability that may be observed within one

2.5 Sampling

Particular attention should be paid to the representativity of samples. Since there is no general procedure applicable to all situations, the sampling programme must be designed for each situation. The location of sampling is usually made at or near the point of discharge to the receiving waterbody, but in the analyses prior to the design of a wastewater treatment, facility samples will be needed from each operation in the fish processing facility. With representativity in mind, samples should be taken more frequently when there is a large variation in flow rate, although wide variations may occur also at constant flow rate.

Commercial sampling equipment is available, although a simple continuous sampler may be constructed with regular tubing and glass carboys as shown in Figure 2.7 (Metcalf and Eddy Inc., 1979).

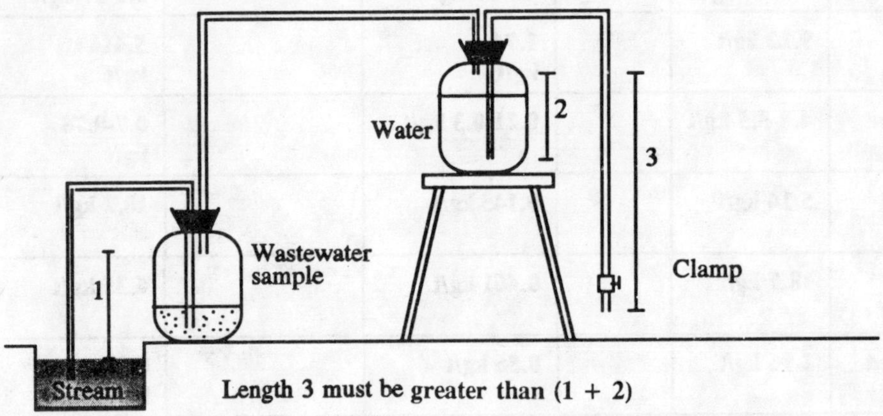

Figure 2.7. Continuous-flow sampling device (Adapted from Metcalf and Eddy Inc.,1979)

The particular procedure for sampling may vary depending on the parameter being monitored. Usually, the description of each standard technique includes recommendation for sampling (Standard Methods, 1989). If possible, samples should be analysed immediately since preservatives often interfere with the tests. In the case of fish processing wastewaters, there is no single method of sample preservation that yields satisfactory results for all cases, and all of them may be inadequate with effluents containing suspended matter. Care should be taken in blending the samples just prior to analysis since in almost all cases they will contain an amount of settleable solids. A case in which the use of preservatives is not recommended is that of the BOD_5. Storage at low temperature (4^oC) may be used with caution for very short periods, and chilled samples should be warmed to 20^oC before analysis. For COD determinations, the samples should be collected in clean glass bottles, and can be preserved by acidification to pH 2 with concentrated sulphuric acid.

For oil and grease, a separate sample should be collected in a wide-mouth glass bottle, well rinsed to remove any trace of detergent. The samples taken for organic nitrogen determination may be preserved similarly for COD. For the determination of solids, it should be checked that the suspended matter does not adhere to the walls, and the samples should be refrigerated at 4^oC to prevent decomposition of biological solids. If phosphorus is to be analysed, preservation can be ensured by storing samples in well rinsed glass bottles (plastic bottles are not recommended) at -10°C and adding 40 mg/l of mercuric chloride.

2.6 Discharge Limits

In order for wastewaters to be discharged into the environment, there are a number of limitations for the concentrations of pollutants but there is no single criterion. In some cases, the limit is established by taking into account where the effluent is to be discharged. This type of criterion is also known as stream standards and is thought to be the more rational, taking into account the assimilative capacity and the intended use of the receiving waterbody.

Another type of criterion is known as effluent standards. In this case, the maximum concentration of a pollutant (e.g., in milligrammes/litre) or the maximum load (e.g., kilogrammes/day) is discharged into a receiving waterbody.

Table 2.3. Summary of discharge limits in the Province of Buenos Aires, Argentina (AGOS, 1990)

Parameter	Unit	Limit if discharged to:		
		Sewage collector	Surface water	Open sea
Temperature	°C	< 45	< 45	< 45
pH		7 to 10	6.5 to 10	6.5 to 10
Settleable solids in 2 h	ml/l	< 5	< 1	< 5
Settleable solids in 10 min.	ml/l	0	0	not spec.
Free chlorine	mg/l	not spec.	< 0.5	< 0.5
BOD_5	mg/l	200	50	150
COD	mg/l	700	250	400
Solubles in ethyl-ether (grease/oil)	mg/1	< 100	< 50	< 100

Examples of these criteria are shown in Table 2.3, where the effluent standards are given in terms of maximum concentration of a pollutant allowed to be discharged to different kinds of waterbodies; and in Table 2.4, where the limits are established for fishery industries as the maximum amount of pollutants to be discharged in a single day, as established by the Environmental Protection Agency in the USA and the proposals by the World Bank.

Table 2.4: Summary of discharge limits proposed by the United States Environmental Protection Agency (EPA), 1985 and the World Bank (WB), 1984

Industry	BOD_5		TSS		Oil/Grease	
	EPA	WB	EPA	WB	EPA	WB
Tuna	20.0	2.2	8.3	2.2	2.1	0.27
Salmon	2.7	11.0	2.6	2.8	0.31	2.8
Other finfish	1.2	4.7	3.1-3.6	4.0	1.0-43	0.85
Crabs	0.3-10	3.6	2.2-19	3.3	0.6-1.8	1.1
Shrimp	63-155	52	110-320	22	36-126	4.6
Clams and oysters	none	41	24-59	41	0.6-2.4	0.62

In all cases, the local authorities should be consulted on the policy to be applied regarding regulations on the volume and concentration of pollutants discharged.

3. PRIMARY TREATMENT

Primary treatment is generally understood as the set of operations performed to remove floatable and settling solids. These solids are present in an effluent prior to secondary treatment in which biological and chemical processes are used to remove most of the remaining organic matter. In primary treatment only physical operations such as screening, sedimentation and flotation are used. The treatment used largely depends on the operations being carried out in the fish processing plant and on the requirements for disposal of the wastewater. Sometimes the only requisite is that no solids settle after 10 minutes, in which case simple screening and/or settling tanks with short residence times may be used. With more stringent norms, more elaborate processes such as flotation and biological treatment will be required.

3.1 Screening

By screening, relatively large solids (0.7 mm or larger) can be removed in a primary treatment facility. This is one of the treatments most commonly used by food processing plants as it quickly reduces the amount of solids being discharged. The simplest configuration is that of flow-through static screens, which have openings of about 1 mm. In some cases, such as streams with fish scales, they may require a scrapping mechanism to minimize clogging. The tangential screens are static but less prone to clogging due to their flow characteristics (Figure 3.1), since the wastewater flow tends to avoid clogging. Removal rates may vary from 40 to 75%.

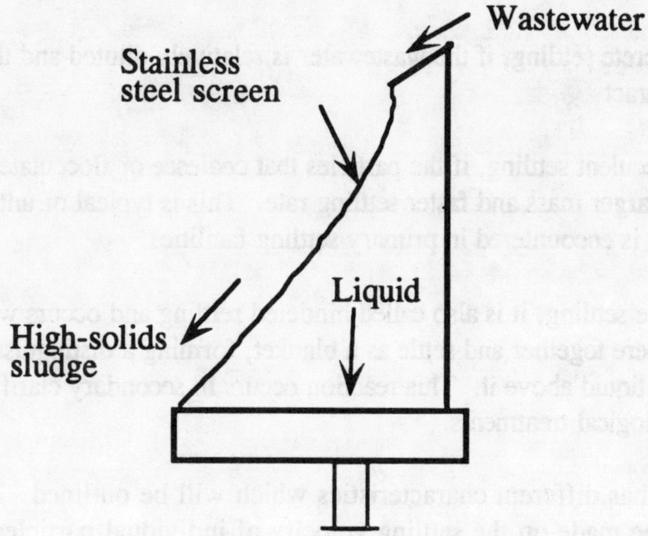

Figure 3.1. Diagram of an inclined or tangential screen

Rotary drum screens have also been used for fisheries wastewaters, although they are mechanically more complex. They consist of a drum which rotates along its axis, and the effluent enters through an opening at one end. Screened wastewater flows outside the drum. The retained solids are washed out from the screen into a collector in the upper part of the drum by a spray of the wastewater.

The screening media used in these devices is generally of stainless material, with openings varying from 0.7 to 1.5 mm. Since the fish solids dissolve in water as time proceeds, it is recommended that the waste streams be screened as soon as possible. By the same token, high intensity agitation of waste streams (such as pumping or flow-through valves) should be minimized before screening or even settling, since they cause breakdown of solids rendering them more difficult to separate.

Both screening and settling are used in fish processing plants, screening being used more frequently in small-scale fish processing plants, together with very simple settling tanks. More elaborated sedimentation devices (discussed in the next section) are used in large-scale factories.

3.2 Sedimentation

Sedimentation is used to remove suspended solids present in the wastewaters. In fisheries wastewaters these include fish scales, portions of fish muscle and offal, the relative proportions varying with the particular process being used.

Sedimentation is based on the difference in density between the bulk of the liquid and the solid particles, which results in the settling of the solids present. The terms sedimentation and settling are often used interchangeably. This operation is conducted not only as part of the primary treatment, but also in the secondary treatment for separation of solids generated in the biological treatments such as activated sludge or trickling filters.

Depending on the properties of solids present in the wastewater, sedimentation can proceed as:

- discrete settling, if the wastewater is relatively diluted and the particles do not interact

- flocculent settling, if the particles that coalesce or flocculate are living particles of larger mass and faster settling rate. This is typical of untreated wastewater and is encountered in primary settling facilities

- zone settling, it is also called hindered settling and occurs when the particles adhere together and settle as a blanket, forming a distinguishable interface with the liquid above it. This reaction occurs in secondary clarifiers for sludges of biological treatments.

Each case has different characteristics which will be outlined. For discrete settling, calculations can be made on the settling velocity of individual particles. In a settling tank, these move both downwards (settling) and towards the outlet zone with the waterflow (see Figure 3.2). The particles that reach the bottom before the outlet will be separated from the effluent while the rest will pass through the settling tank. The critical velocity (v_c) below in which a particle will be carried out of the tank is given by the depth of liquid (d), the volume of the tank (V) and the flow rate (Q):

$$v_c = d / (V/Q)$$

The ratio of V/Q is also known as the residence time of the liquid in the tank. It is called the overflow rate when v_c is expressed in terms of volume of effluent per unit surface area of the tank per unit of time. This case may be present in fisheries wastewaters but is not the most common.

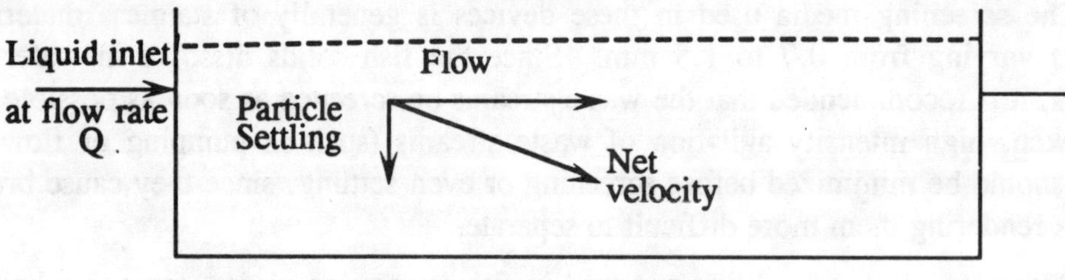

Figure 3.2. Schematics of discrete settling

In the case of a flocculant suspension, the formation of larger particles due to coalescence depends on several factors, e.g., the nature of the particles and the rate of coalescence. A theoretical analysis is not feasible due to the interaction of particles which depends among other factors on the overflow rate, the concentration of particles and the depth of the tank.

A settling column is used to evaluate the settling characteristics of a flocculant suspension (see Figure 3.3). The same kind of column using only one sampling port can be used to study the discrete settling.

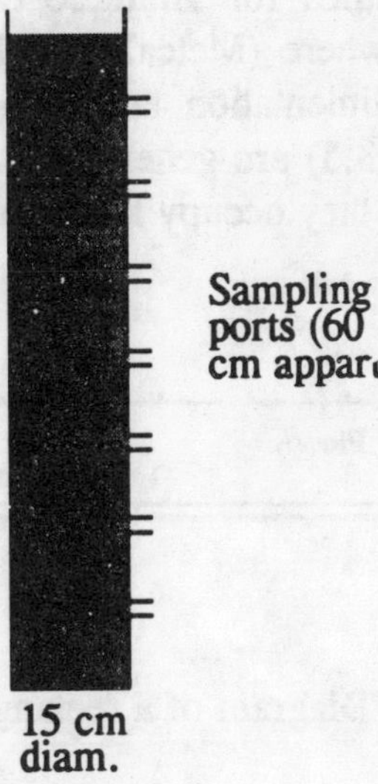

Figure 3.3. Laboratory settling column

The zone (or hindered) settling, which occurs when the particles do not settle independently, is also studied by batch tests. In this case an effluent that is initially uniform in solids concentration (see Figure 3.4), if allowed to settle, will do so in zones, the first of which is that of clarified water (1), below is the interfacial zone (2) in which the solids concentration is considered uniform. In the bottom a compact sludge develops in the so called compaction zone (4). Between (2) and (4), a transition zone (3) generally exists.

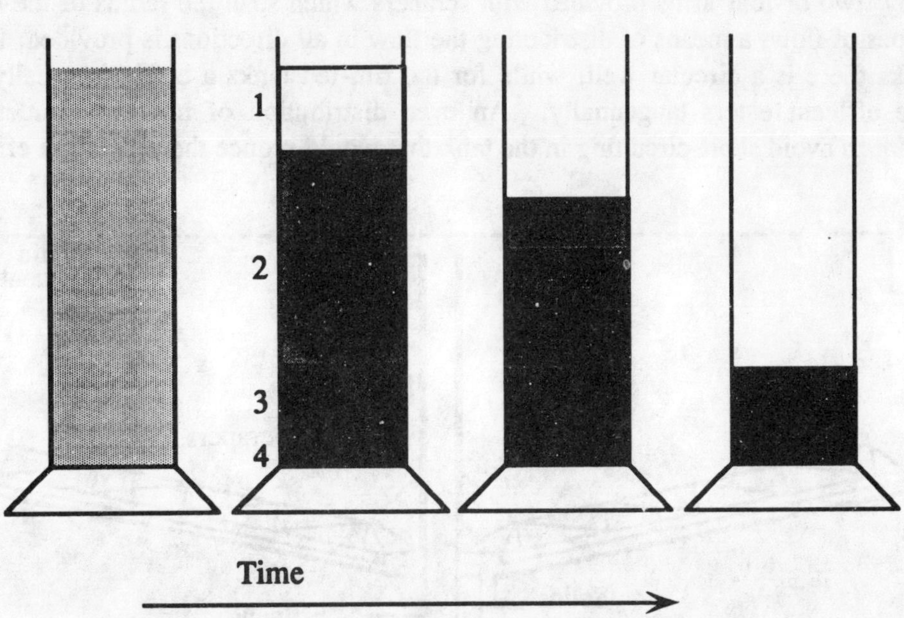

Figure 3.4. Diagram of a zone settling process

As time proceeds, the clarified effluent and compaction zones will increase in size while the two intermediates will eventually disappear. In some cases, further compaction may occur.

The detailed design procedures for all these cases are beyond the scope of this document, and can be found elsewhere (Metcalf and Eddy Inc., 1979; Ramalho, 1977). The actual configuration of a sedimentation tank can be either rectangular or circular. Rectangular settling tanks (Figure 3.5) are generally used when several tanks are required and there is space constraint, since they occupy less space than several circular tanks.

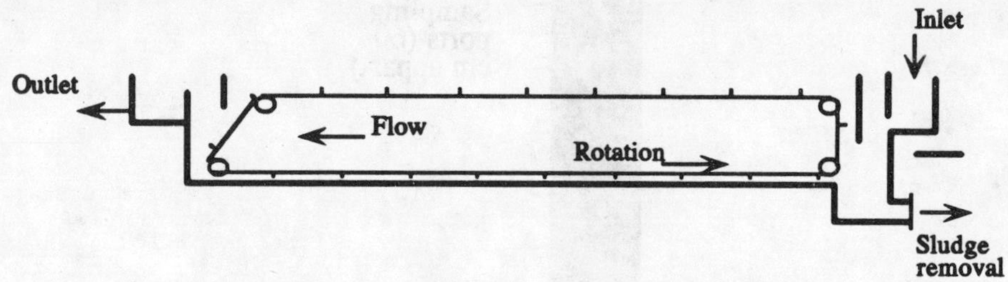

Figure 3.5. Diagram of a rectangular clarifier

For removal of solids, a series of chain-driven scrapers are used: these span the width of the tank, are regularly spaced and move at 0.5 to 1 m/min. The sludge is collected in a hopper in the end of the tank, where it may be removed by screw conveyors or pumped out. Configurations exist in which the sludge is forced opposite to the flow, as shown here, but concurrent flow of the liquid and solids is also used.

The circular tanks are reported to be more effective. In these, the effluent circulates radially, the water being introduced at the periphery or from the centre. Figure 3.6 shows such a configuration. The solids removal are generally removed from near the centre, for which a slope of 10% is required in the bottom of the tank. The sludge is forced to the outlet by two or four arms provided with scrapers which span the radius of the tank. For both types of flow, a means of distributing the flow in all directions is provided: for centre-fed tanks there is a circular well, while for the rim-fed tanks a baffle is usually installed and the effluent enters tangentially. An even distribution of inlet and outlet flows is important to avoid short-circuiting in the tank that would reduce the separation efficiency.

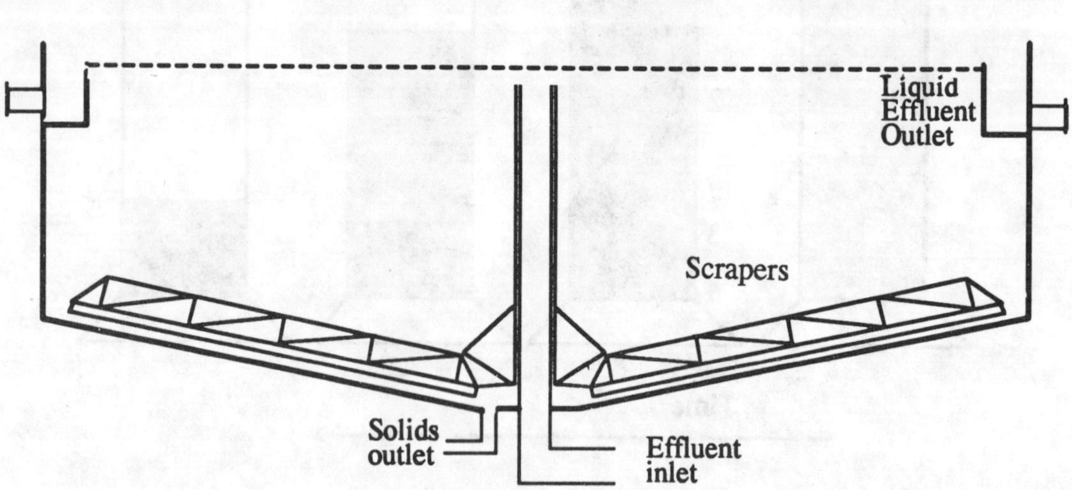

Figure 3.6. Diagram of a rectangular clarifier with centre feed

A critical factor for selection of tank size is the so-called surface-loading rate, generally expressed as volume of wastewater per unit time and unit area of settler (m^3/m^2d). This loading rate depends on the characteristics of the effluent and the solids content, and can be determined from the settling tests described above. The retention time in the settlers is generally in the order of one to two hours, but the capacity of the tanks must be determined taking into account the peak flow rates so that good separation is also obtained in these cases.

Some settling tanks, especially the larger ones, are provided with a mechanism for scum removal, since in biological wastes such as fisheries wastewaters, its formation is almost unavoidable.

In cases of small or elementary settling basins, the sludges can be removed using an arrangement of perforated piping placed in the bottom of the settling tank (Rich, 1980). The pipes must be placed regularly spaced (Figure 3.7), be of a diameter wide enough to be cleaned easily in case of clogging and the flow velocities should also be high enough to prevent sedimentation. These last two requisites are somewhat contradictory and a compromise is usually reached, using pipes of 5 cm in diameter, perforated with holes of 1-1.5 cm in diameter, 1 m apart. Flow in individual pipes may be regulated by valves. This configuration is best used after screening and is also found in biological treatment tanks for sludge removal.

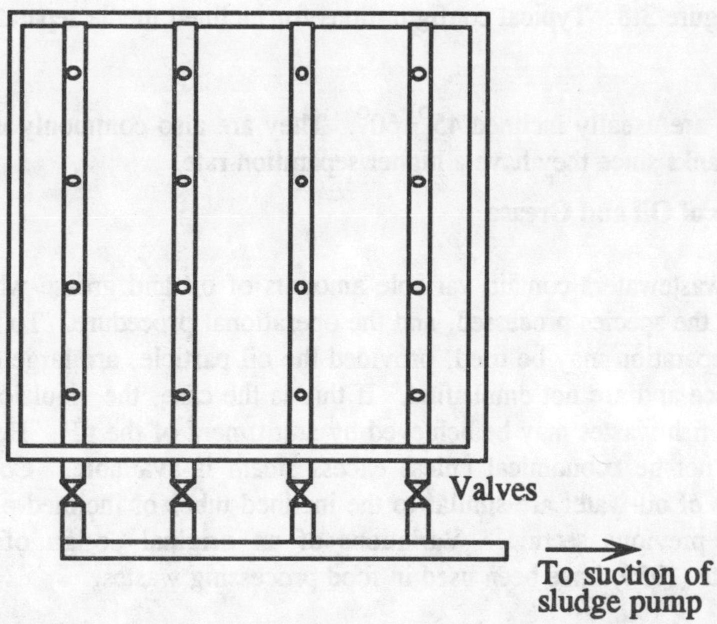

Figure 3.7. Pipe arrangement for sludge removal from settling tanks

An alternative to the above configurations for settling tanks is that of the inclined tube separators (Hansen and Culp, 1967). These separators consist of tubes (although there are alternate designs that use plates close to each other) which are tilted (see Figure 3.8).

The concept is that, when a settling particle reaches the wall of the tube or the lower plate, it coalesces with another particle to give one of larger mass and higher settling rate.

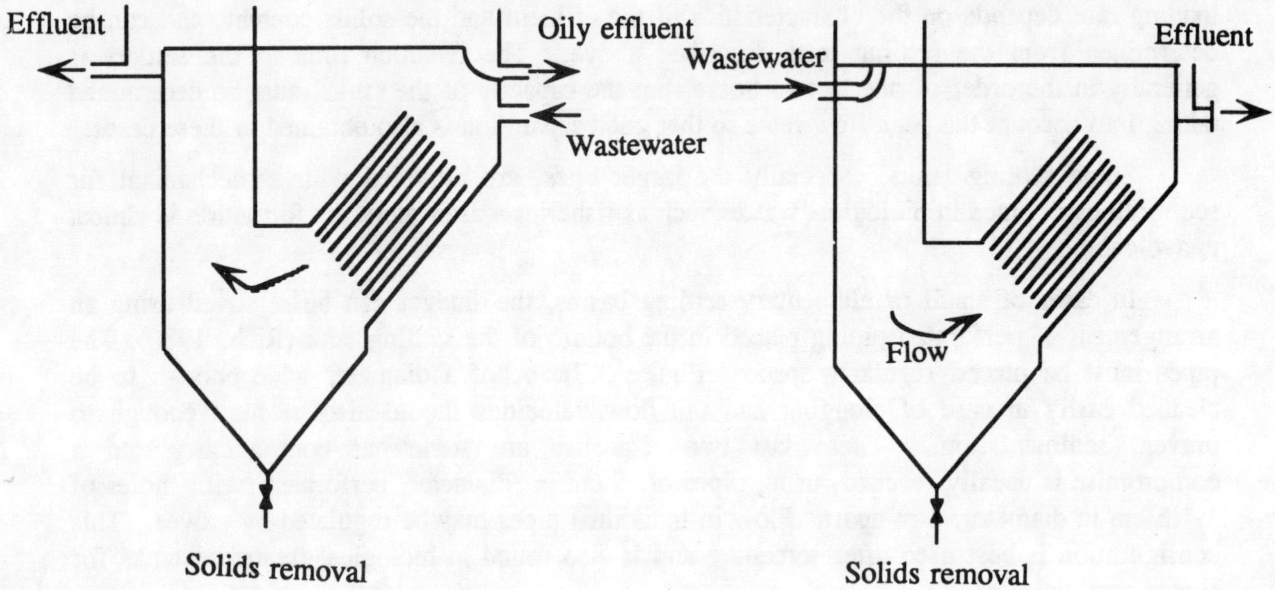

Figure 3.8. Typical configurations for inclined media separators

The media are usually inclined $45^o - 60^o$. They are also commonly used to upgrade existing settling tanks since they have a higher separation rate.

3.3 Separation of Oil and Grease

Fisheries wastewaters contain variable amounts of oil and grease which depend on the process used, the species processed, and the operational procedure. To remove oil and grease, gravity separation may be used, provided the oil particles are large enough to float towards the surface and are not emulsified. If this is the case, the emulsion must be first broken, which in fish wastes may be achieved by adjustment of the pH. Heat may also be used but it may not be economical unless excess steam is available. Configurations of gravity separators of oil-water are similar to the inclined tubes or inclined plates separators discussed in the previous section. Variations of an original design of the American Petroleum Institute (1959) have been used in food processing wastes.

3.4 Flotation

Flotation is an operation that removes not only oil and grease but also suspended solids. It is discussed in this section since it is one of the most effective systems for suspensions which contain oil and grease. The most common procedure is that of dissolved air flotation (DAF), in which the waste stream is first pressurized with air in a closed tank. After passing through a pressure-reduction valve, the wastewater enters the flotation tank

(Figure 3.9a) where, due to the sudden reduction in pressure, minute air bubbles in the order of 50-100 microns in diameter are formed. As the bubbles rise to the surface, the suspended solids and oil or grease particles adhere to them and are carried upwards. It is common practice to use chemicals to enhance flotation performance. As with coagulants (discussed later) these aids should preferably be innocuous, since these recovered solids are frequently used in animal feed formulations.

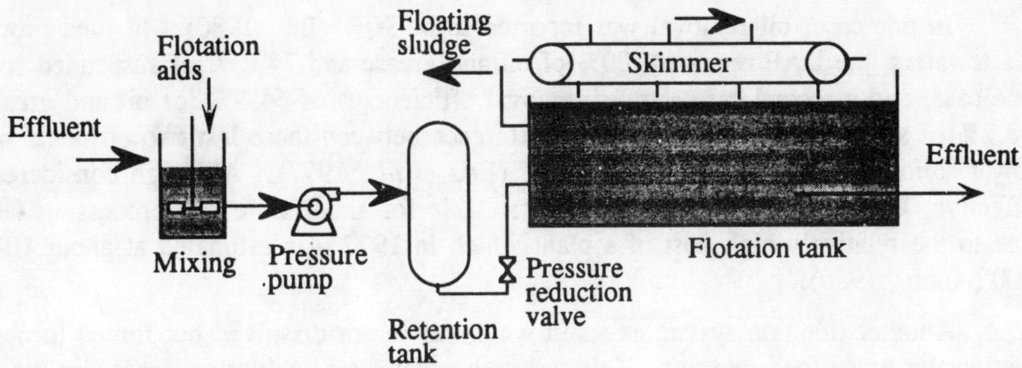

Figure 3.9a. Diagram of a dissolved air flotation system

One alternate design involves the recycling of part (10-30%) of the treated water (Figure 3.9b). All systems contain a mechanism for removing the solids that may settle to the bottom of the flotation tanks, usually by a helical conveyor placed in the conical bottom. The main advantage claimed of DAF systems is the faster rate at which very small or light suspended solids can be removed in comparison with settling.

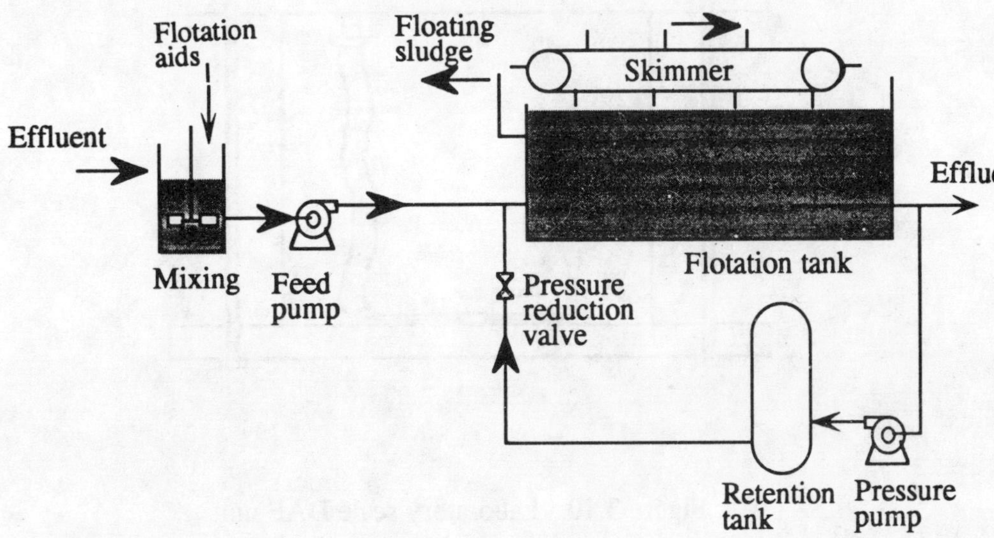

Figure 3.9b. Diagram of a DAF system with recycle

Performance of DAF systems has been reported to be dependent on several factors, of which one of the most important is the solids concentration; higher solids content usually gives higher removal efficiencies. Other factors affecting the efficiency of the operation are the ratio of air to solids (A/S) defined as the amount of air released after pressure reduction and the amount of solids present in the wastewater. There is usually an optimum A/S which is determined by bench scale tests.

Key factors in the successful operation of DAF units are the maintenance of proper pH (usually between 4.5 and 6, with 5 being most common to minimize protein solubility and break-up emulsions), proper flow rates and the continuous presence of trained operators.

In one case, oil removal was reported to be 90% (Ilet, 1980). In tuna processing wastewaters, the DAF removed 80% of oil and grease and 74.8 % of suspended solids in one case, and a second case showed removal efficiencies of 64.3% for oil and grease and 48.2% of suspended solids. The main difference between these last two effluents was the lower solids usually content of the second (Ertz *et al.*, 1977). Although considered very effective, DAF systems are probably not suitable for small-scale fish processing facilities due to the relatively high cost of a plant which in 1977 was estimated at about US$ 250 000 (Anon., 1986).

Another flotation system exists in which air is not dissolved but forced through the wastewater by surface aerators. This system generates air bubbles of larger size than DAF systems and no report exists about its application to fisheries wastewaters.

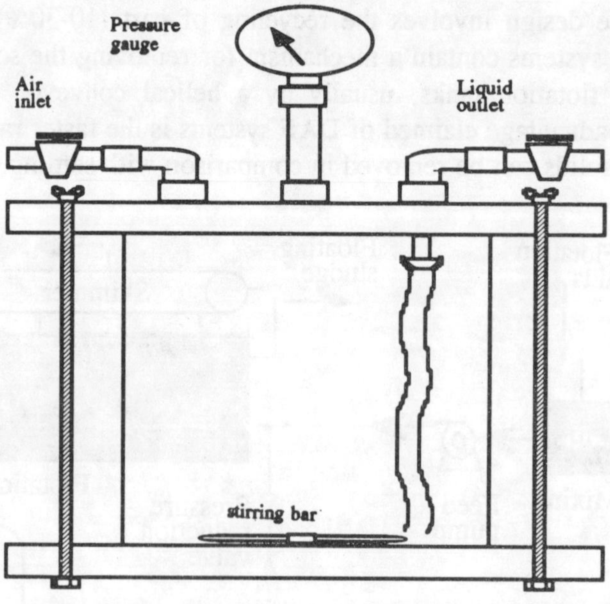

Figure 3.10. Laboratory scale DAF unit

Prior to the design or selection of a DAF system, it is advisable to carry out laboratory experiments to evaluate its applicability and critical operating factors such as the air to solids ratio, the effectiveness of flocculants and the proper pH. This can be conveniently done in laboratory units such as that shown in Figure 3.10. In these devices, water with or without chemicals and pH adjustment is introduced, and the pressure raised to the desired value. After mixing to saturate the liquid with air, pressure is released and the liquid flows to a graduated cylinder where time is allowed for separation. The detailed procedures for conducting the evaluations are available elsewhere (Eckenfelder and Ford, 1969).

4. BIOLOGICAL TREATMENT

The goal of all biological wastewater treatment systems is to remove the non-settling solids and the dissolved organic load from the effluents by using microbial populations. Biological treatments are generally part of secondary treatment systems. The micro-organisms used are responsible for the degradation of the organic matter and the stabilization of organic wastes. With regard to the way in which they utilize oxygen, they can be classified into aerobic (require oxygen for their metabolism), anaerobic (grow in absence of oxygen) and facultative (can proliferate either in absence or presence of oxygen although using different metabolic processes). Most of the micro-organisms present in wastewater treatment systems use the organic content of the wastewater as an energy source to grow, and are thus classified as heterotrophes from a nutritional point of view. The population active in a biological wastewater treatment is mixed, complex and interrelated. By example, in a single aerobic system, members of the genera *Pseudomonas, Nocardia, Flavobacterium, Achromobacter and Zooglea* may be present, together with filamentous organisms (*Beggioata* and *Spaerotilus* among others). In a well-functioning system, protozoas and rotifers are usually present and are useful in consuming dispersed bacteria or non-settling particles. More extensive description and treatment of the microbiology of wastewater treatment systems are given elsewhere (Stanier, 1976).

The organic load present is incorporated in part as biomass by the microbial populations, and almost all the rest is liberated as gas (carbon dioxide (CO_2) if the treatment is aerobic, or carbon dioxide plus methane (CH_4) if the process is anaerobic) and water. In fisheries wastewaters the non-biodegradable portion is very low.

Unless the cell mass formed during the biological treatment is removed from the wastewater (e.g., by sedimentation or other treatment described in the previous section), the treatment is largely incomplete, because the biomass itself will appear as organic load in the effluent and the only pollution reduction accomplished is that fraction liberated as gases.

The biological treatment processes used for wastewater treatment are broadly classified as aerobic (in which aerobic and facultative micro-organisms predominate) or anaerobic (which use anaerobic micro-organisms).

If the micro-organisms are suspended in the wastewater during biological operation, the operations are "called suspended growth processes", while the micro-organisms that are attached to a surface over which they grow are called "attached growth processes".

This section explains the principles and main characteristics of the most common processes in each case.

4.1 Aerobic Processes

In these, the reactions occurring can be summarized as:

$$\text{organic load} + \text{oxygen} + \text{more cells} + CO_2 + H_2O$$

In fisheries wastewaters, the need for addition of nutrients (the most common being nitrogen and phosphorus) seldom appears, but an adequate provision of oxygen is essential for successful operation of the systems.

The most common aerobic processes are: activated sludge systems, lagoons, trickling filters and rotating disk contactors. These aerobic processes are described, together with the devices used for aeration.

4.1.1 Activated Sludge Systems

These systems originated in England in the early 1900's and earned their name because a sludge (mass of microbes) is produced which aerobically degrades and stabilizes the organic load of a wastewater. Figure 4.1 shows the layout of a typical activated sludge system.

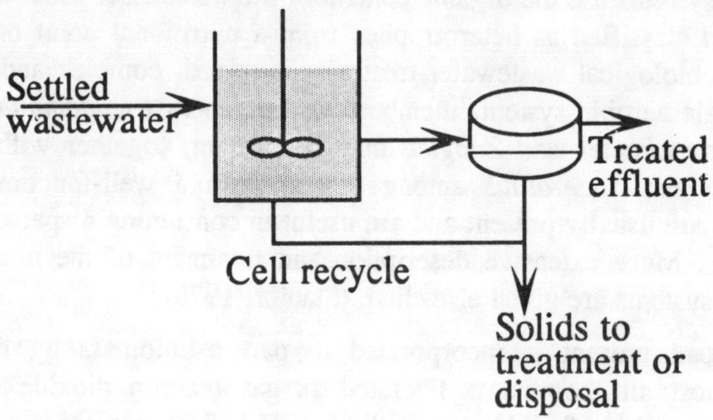

Figure 4.1. Diagram of a simple activated sludge system

For larger systems, especially when high variability is expected, the design involves the use of multiple aeration tanks and multiple settling tanks. The number of units employed depends on the flow of wastewater being generated.

The organic load (generally coming from primary treatment operations such as settling, screening or flotation) enters the reactor where the active microbial population (activated sludge) is present. The reactor must be continuously aerated. The mixture then passes to a secondary settling tank where the cells are settled. The treated wastewater is generally discharged after disinfection while the settled biomass is recycled in part to the aeration basin. The cells must be recycled in order to maintain sufficient biomass to degrade the organic load as quickly as possible. The amount that is recirculated depends on the need to obtain a high degradation rate and on the need for the bacteria to flocculate properly so that the secondary settling separates the cells satisfactorily. As the cells are retained longer in the system, the flocculating characteristics of the cells improve since they start to produce extra cellular slime which favours flocculating.

The most common types of activated sludge are the conventional and the continuous flow stirred tank (Figure 4.1), in which the contents are completely mixed. In the conventional process, the wastewater is circulated along the aeration tank, with the flow being arranged by baffles in plug flow mode (Figure 4.2). The oxygen demand for this arrangement is maximum at the inlet, as is the organic load concentration.

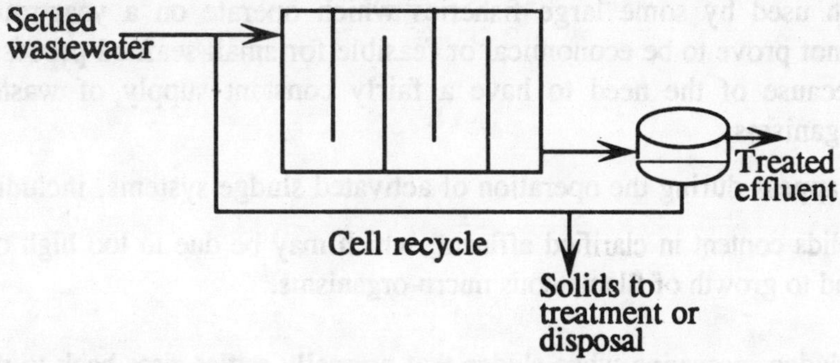

Figure 4.2: Diagram of a conventional activated sludge process

In the completely mixed process the inflow streams are usually introduced at several points to facilitate the homogeneity of the mixing; if the mixing is complete, the properties are constant throughout the reactor. This configuration is inherently more stable to perturbations because mixing causes the dilution of the incoming stream into the tank. In fisheries wastewaters the perturbations that may appear are peaks of concentration of organic load or flow peaks. The flow peaks can be damped in the primary treatment tanks. The conventional configurations would require less reactor volume if smooth plug flow could be assured, which usually does not occur.

Other versions of activated sludge systems (e.g., extended aeration, contact stabilization, step aeration and pure oxygen processes) are used in other kinds of wastewaters but are not known to be applied to treat fisheries wastewaters. They are discussed elsewhere (Metcalf and Eddy Inc., 1979; Eckenfelder, 1980).

In all activated sludge systems, the cells are separated from the liquid and partially returned to the system to have a relatively high concentration of cells that degrade the organic load in a relatively short time. Therefore, two different resident times are characteristic: the hydraulic residence time (θ_H) given by the ratio of reactor volume (V) to flow of wastewater (Q):

$$\theta_H = V / Q$$

and the cell residence time (θ_C) given by the ratio of cells present in the reactor to the mass of cells wasted per day. Typical θ_H values are in the order of 3-6 hours, while θ_C fluctuates between 3 and 15 days. Such difference in resident times is obtained by discharging the clarified effluent but wasting only a small fraction of the sludge. This in turn can be accomplished by discarding a portion of the sludge from the settling tank or by wasting a fraction of the outlet of the reactor before entering the settling tank.

In activated sludge systems, organic load removals of 85-95% are the most common. A key factor in the success of these systems is its proper operation, which requires trained manpower. Although used by some large fisheries which operate on a year-round basis, activated sludge may not prove to be economical or feasible for small seafood processors who operate seasonally because of the need to have a fairly constant supply of wastewater to maintain the micro-organisms.

Problems may appear during the operation of activated sludge systems, including :

- High solids content in clarified effluent, which may be due to too high or too low solids retention time and to growth of filamentous micro-organisms.

- Rising sludge, occurring when sludge that normally settles rises back to the surface after having settled. In most cases, this is caused by the denitrification process, where nitrate present in the effluent is reduced to nitrogen gas, which then becomes trapped in the sludge causing this to float. This problem can be reduced by decreasing the flow from the aeration basin to the settling tank or reducing the sludge resident time in the settler, either by increasing the rate of recycle to the aeration basin, increasing the rate of sludge collection from the bottom, or increasing the sludge wasting rate from the system.

- Bulking sludge, that which settles too slowly and is not compactable, caused by the predominance of filamentous organisms. This problem can be due to several factors of which the most common are nutrient balance, wide fluctuations in organic load, oxygen limitation (too low levels), and an improper sludge recycle rate.

- Insufficient reduction of organic load, probably caused by a low solids retention time, insufficient amount of nutrients such as P or N (rare in fisheries wastewaters), short-circuiting in the settling tank, poor mixing in the reactor and insufficient aeration or presence of toxic substances.

- Odours, caused by anaerobic conditions in the settling tanks or insufficient aeration in the reactor.

4.1.2 Aerated lagoons

The aerated lagoons are basins, normally excavated in earth and operated without solids recycling into the system. This is the major difference with respect to activated sludge systems. Two types are the most common: the completely mixed lagoon (also called completely suspended) in which the concentration of solids and dissolved oxygen are maintained fairly uniform and neither the incoming solids nor the biomass of microorganisms settle, and the facultative (aerobic-anaerobic or partially suspended) lagoons. In the facultative lagoons, the power input is reduced causing accumulation of solids in the bottom which undergo anaerobic decomposition, while the upper portions are maintained aerobic (Figure 4.3 gives an example). The main operational difference between these lagoons is the power input, which is in the order of 2.5-6 Watts per cubic metre (W/m^3) for aerobic lagoons while the requirements for facultative lagoons are of 0.8-1 W/m^3. Being open to the atmosphere, the lagoons are exposed to low temperatures which can cause reduced biological activity and eventually the formation of ice. This can be partially alleviated by increasing the depth of the basin. These units require a secondary sedimentation unit, which in some cases can be a shallow basin excavated in earth, or onventional settling tanks can be used.

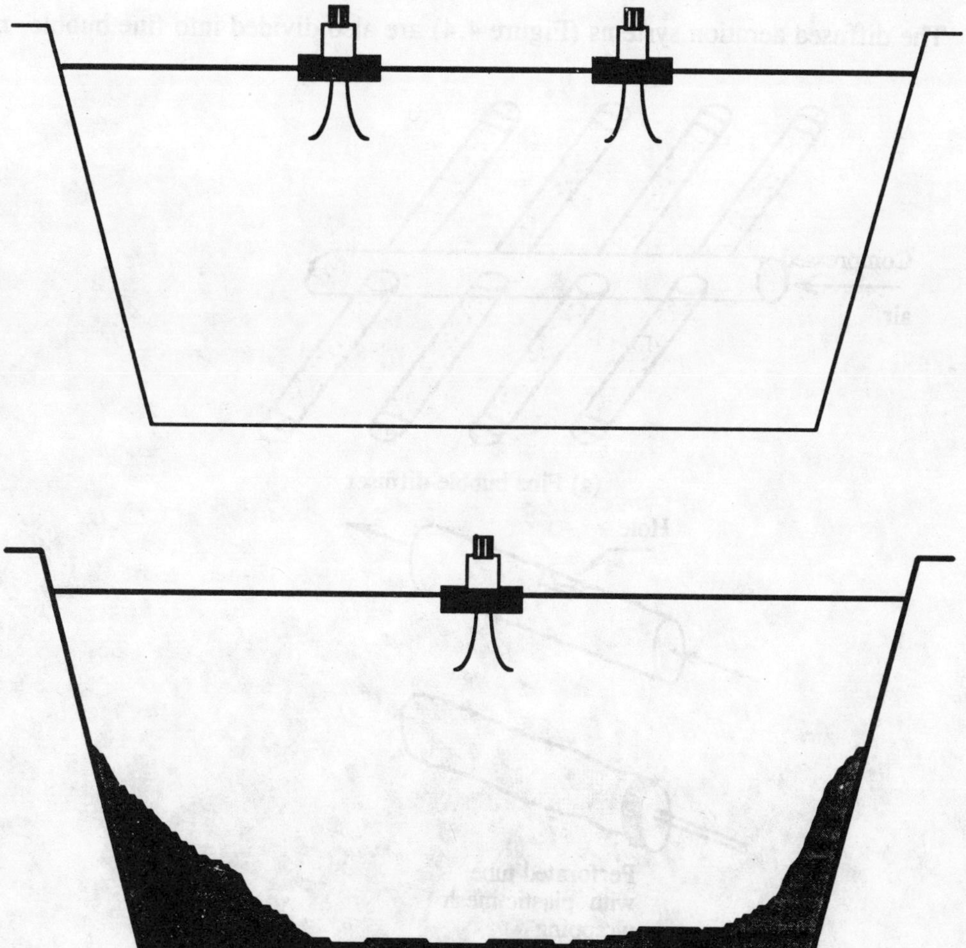

Figure 4.3. Diagram of aerobic (top) and facultative (bottom) aerated lagoons

If excavated basins are used for settling, care should be taken to provide a residence time long enough for the solids to settle, and there should also be provision for the accumulation of sludge. There is a very high possibility of offensive odour development due to the decomposition of the settled sludge, and algae might develop in the upper layers contributing to an increased content of suspended solids in the effluent. Odours can be minimized by using minimum depths of up to 2m, while algae production is reduced with liquid retention time of less than two days.

The solids will also accumulate, all along the aeration basins in the facultative lagoons and even in corners, or between aeration units in the completely mixed lagoon. These accumulated solids will, on the whole, decompose in the bottom, but since there is always a non-biodegradable fraction, a permanent deposit will build up. Therefore, periodic removal of these accumulated solids becomes necessary.

4.1.3 Aeration

The aerated systems described above need an oxygen supply. Depending on the characteristics of the process, different designs may be used. The oxygen can be supplied to the activated sludge by either diffused aeration, by turbine agitation, by static aerators, or by surface coarse or large bubble diffusers. The last two are used also in the lagoon systems.

The diffused aeration systems (Figure 4.4) are also divided into fine bubble, medium

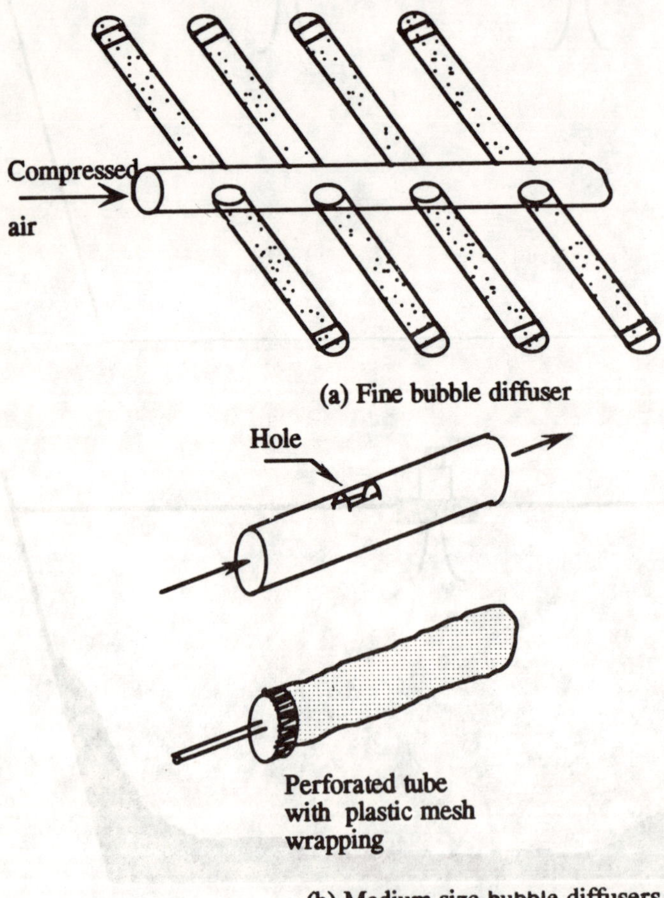

(a) Fine bubble diffuser

(b) Medium size bubble diffusers

Figure 4.4. Fine and medium bubble diffusers

and coarse or large bubble diffusers. The fine bubble diffusers are built of porous materials (grains of pure silica or aluminum oxide are bonded ceramically or by resins) which provide very small bubbles of high surface area that favour the oxygen transfer from the air to the wastewater. The medium bubble diffusers are perforated pipes or tubes wrapped with plastic or woven fabric. The coarse or large bubble diffusers can be orifice devices of various types, some of which are designed to be non-clogging (Figure 4.5).

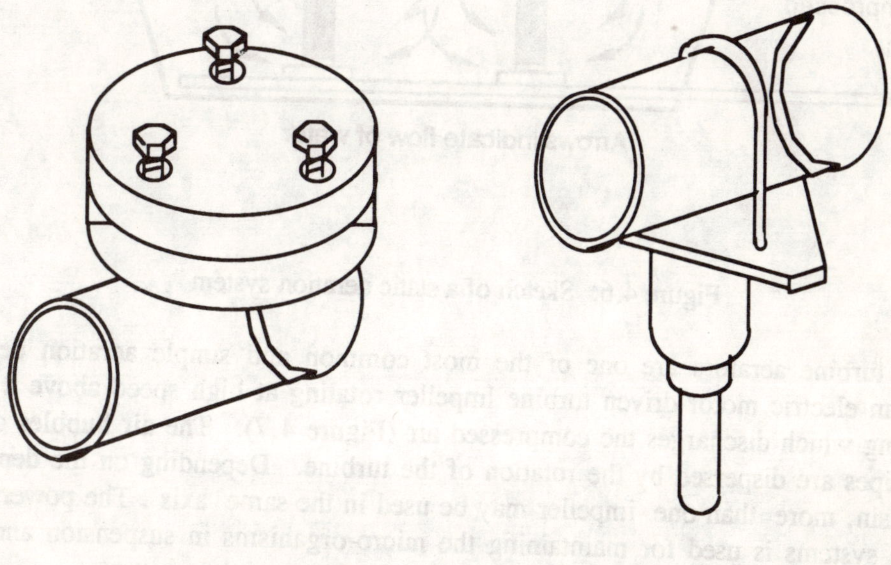

Figure 4.5. Coarse or large bubble diffusers

With the small or fine bubble diffusers, it is important to use air free of particles that would otherwise clog them. Although somewhat less efficient for oxygen transfer, the coarse bubble diffusers are sometimes preferred because the presence of particles in the air is not a critical problem, and also for their lower cost and maintenance requirements. The diffusers are placed along air manifolds, close to the bottom of the aeration tanks.

The static aerators (Figure 4.6) are vertical tubes placed at the bottom of the aeration tank, with packing material along its length. The compressed air is supplied from the bottom of the tubes, forcing a mixture of air and water through the packing, where most of the oxygen transfer to the wastewater takes place. They have been used mainly in aerated lagoons.

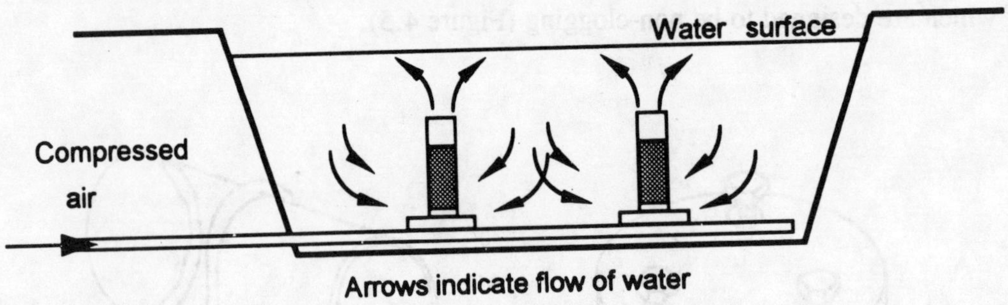

Figure 4.6. Sketch of a static aeration system

The turbine aerators are one of the most common and simple aeration devices and consist of an electric motor-driven turbine impeller rotating at high speed above a pipe or a sparging ring which discharges the compressed air (Figure 4.7). The air bubbles discharged from the pipes are dispersed by the rotation of the turbine. Depending on the depth of the aeration basin, more than one impeller may be used in the same axis . The power drawn by the turbine systems is used for maintaining the micro-organisms in suspension and to break down and disperse the air bubbles, the latter demanding most of the power.

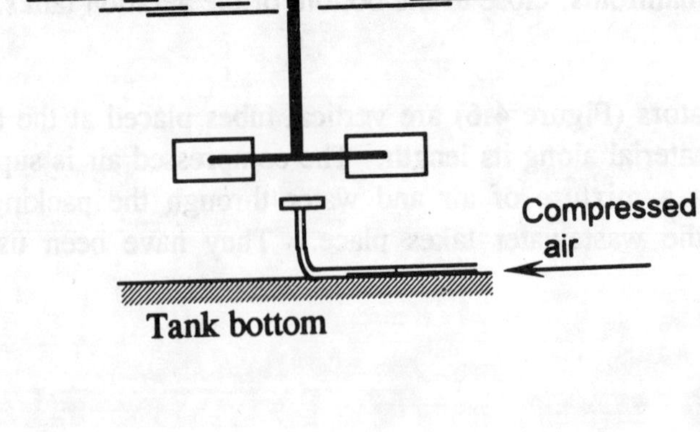

Figure 4.7. Turbine aeration system

The most common surface aeration units (Figure 4.8) are mounted on a float and consist of a propeller installed inside a rising tube and driven by a non-immersed motor. The propeller draws the liquid from under the unit and sprays it above the surface of the tank. The oxygen transfer takes place from the air to the droplets sprayed and to the turbulent surface of the liquid surrounding these units. Other surface aeration units are the so-called "brush" aerators which are basically blades mounted on a cylinder which rotates through the liquid (Figure 4.9). Usually, these units require baffles to direct the flow and insure turbulent velocity.

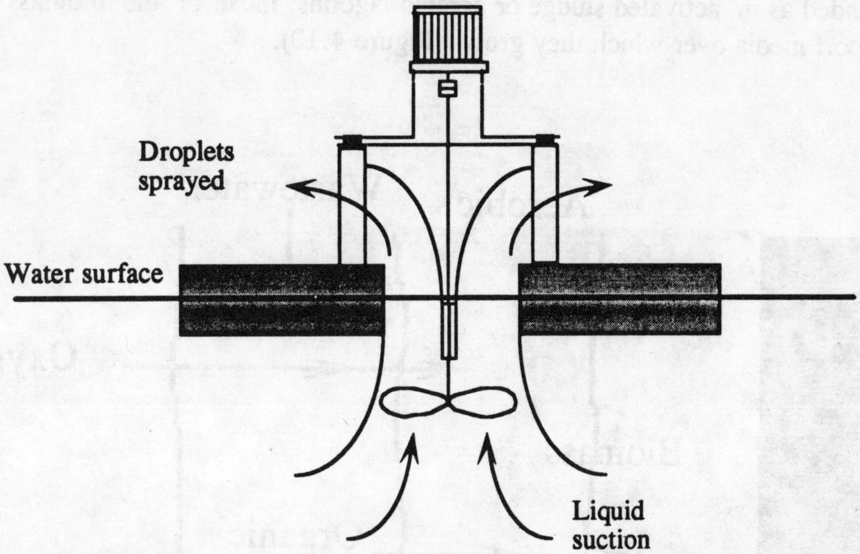

Figure 4.8. Diagram of a floating surface aerator

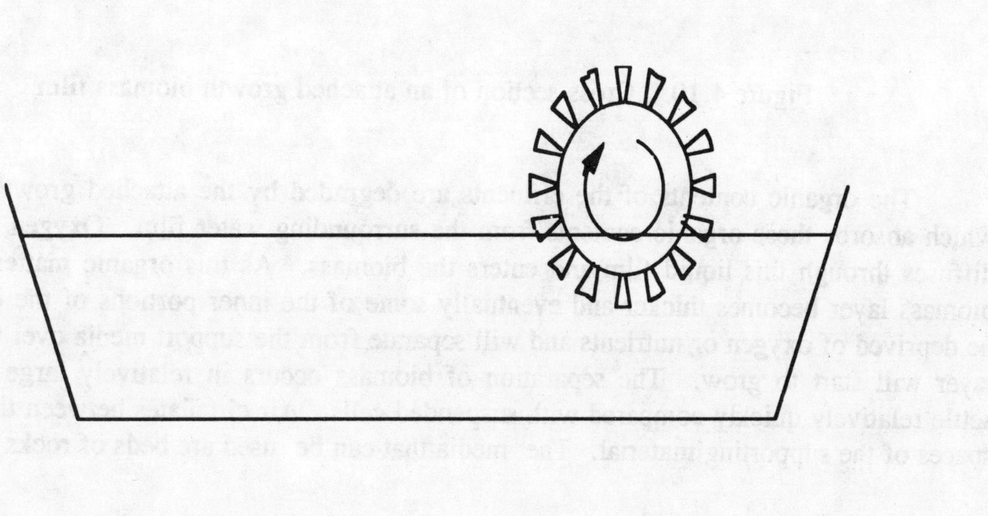

Figure 4.9. Sketch of a surface brush aerator

The oxygen transfer rate of the different devices fluctuate between 0.7 and 1.4 kg of oxygen per kiloWatt-hour when used in actual wastewaters. Most catalogues give much higher transfer capacities, because these values are based on test under standard conditions (typically clean, tap water at 20°C and no dissolved oxygen at the start of the test). When selecting aeration equipment, care should be taken in interpreting these values and transfer rates in actual wastewater should be requested for proper evaluation.

4.1.4 Trickling filters

The trickling filter is one of the most common attached growth processes. Rather than being suspended as in activated sludge or aerated lagoons, most of the biomass is attached to some support media over which they grow (Figure 4.10).

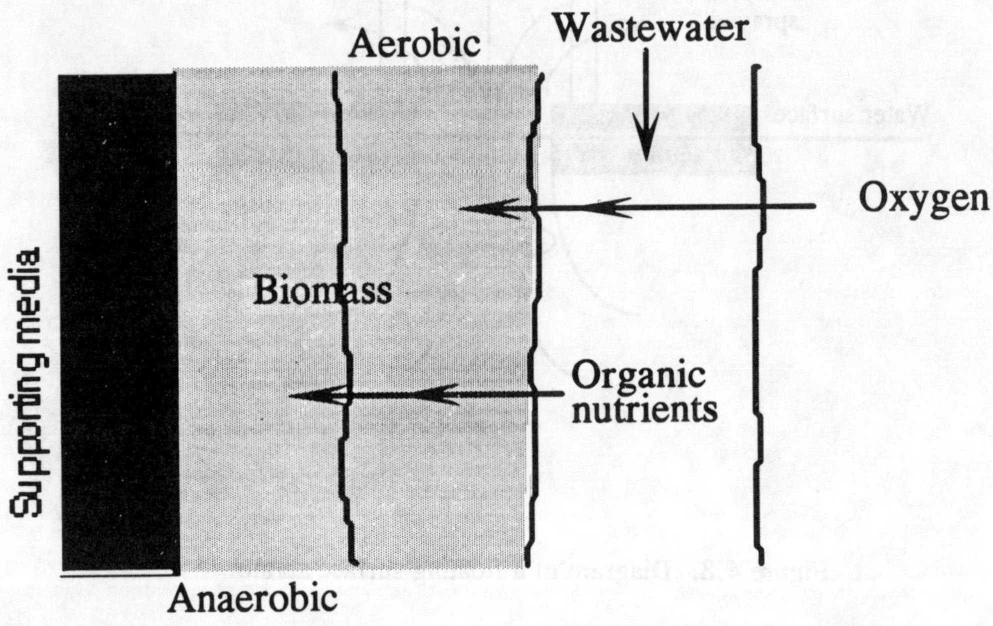

Figure 4.10. Cross-section of an attached growth biomass film

The organic contents of the effluents are degraded by the attached growth population which absorbs these organic contents from the surrounding water film. Oxygen from the air diffuses through this liquid film and enters the biomass. As this organic matter grows, the biomass layer becomes thicker and eventually some of the inner portions of the biomass will be deprived of oxygen or nutrients and will separate from the support media over which a new layer will start to grow. The separation of biomass occurs in relatively large flocs which settle relatively quickly compared with suspended cells. Air circulates between the interstitial spaces of the supporting material. The media that can be used are beds of rocks (ranging in size from 5 to 10 cm) randomly packed, although regular packings of plastic material (Figure 4.11) are becoming more common recently in view of its much lighter weight, better flow distribution, larger void space and specific area.

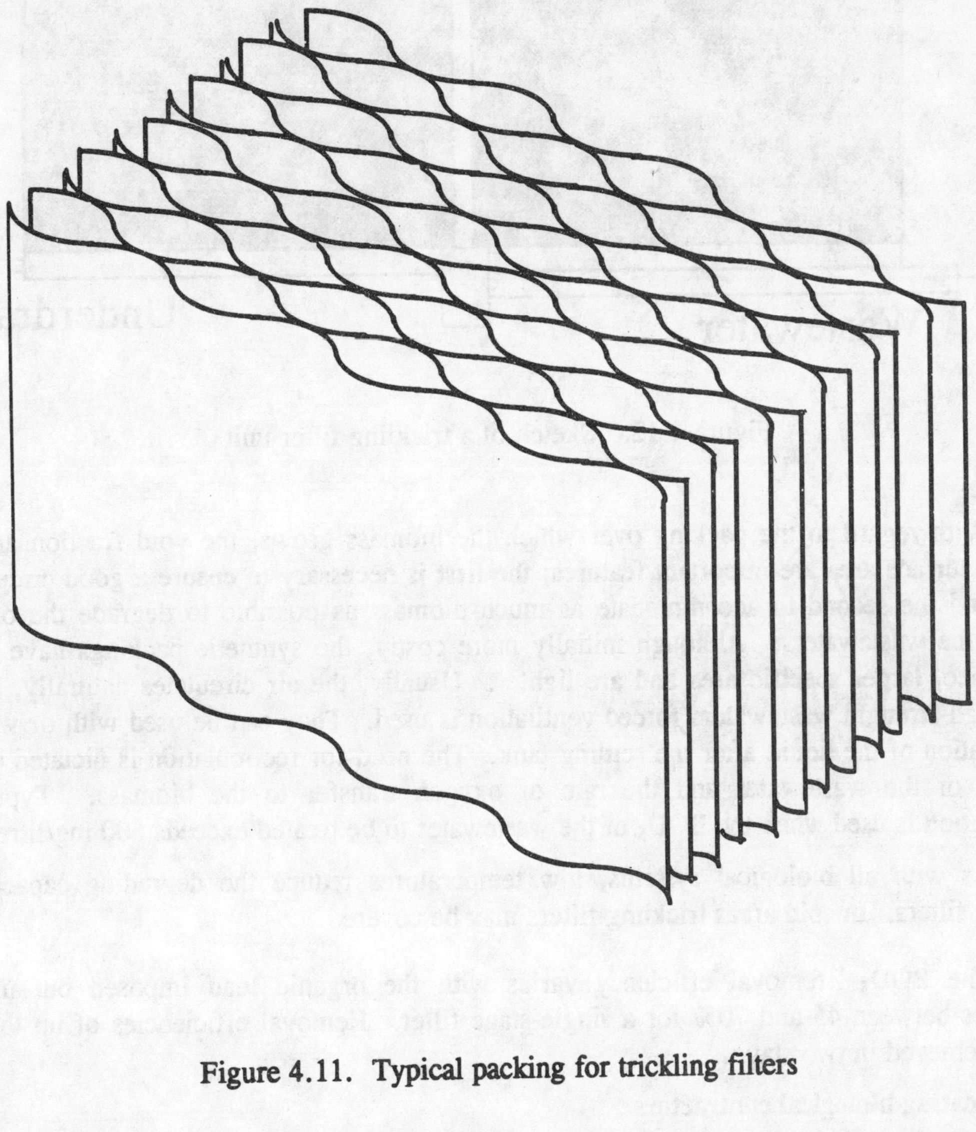

Figure 4.11. Typical packing for trickling filters

The trickling filter units consist of a circular tank filled with the packing media in depths from 1 to 2.5 m, or 10 m if synthetic packing is used. The bottom of the tank must be constructed rigid enough to support the packing and also designed to collect the treated wastewater which is either sprayed by regularly-spaced nozzles or (more common) by rotating distribution arms (Figure 4.12). The liquid percolates through the packing and the organic load is absorbed and degraded by the biomass while the liquid drains to the bottom where it is collected.

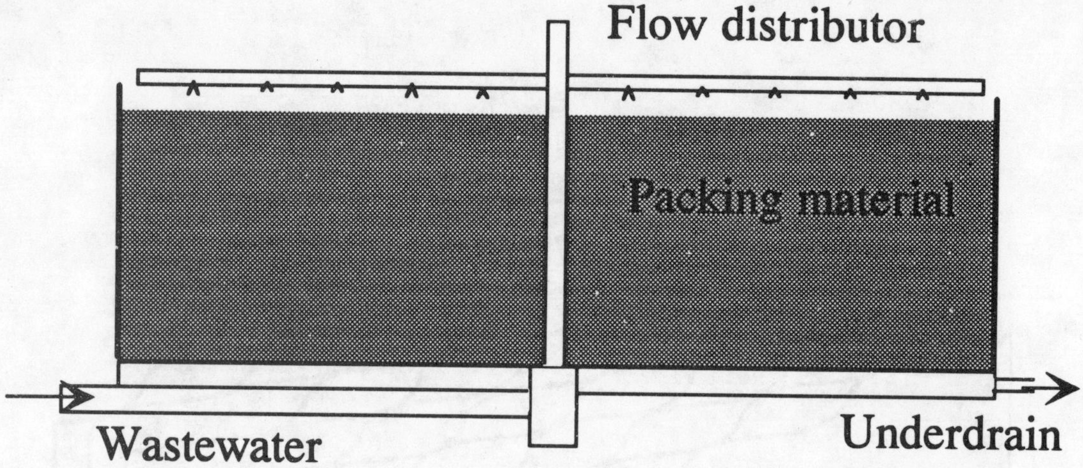

Figure 4.12. Sketch of a trickling filter unit

With regard to the packing over which the biomass grows, the void fraction and the specific surface area are important features; the first is necessary to ensure a good circulation of air and the second to accommodate as much biomass as possible to degrade the organic load of the wastewaters. Although initially more costly, the synthetic packings have larger void space, larger specific area and are lighter. Usually, the air circulates naturally, but in some high-strength wastewaters forced ventilation is used. They can be used with or without recirculation of the liquid after the settling tank. The need for recirculation is dictated by the strength of the wastewater and the rate of oxygen transfer to the biomass. Typically, recirculation is used when the BOD_5 of the wastewater to be treated exceeds 500 mg/litre.

As with all biological systems, low temperatures reduce the degrading capacity of trickling filters. In cold areas trickling filters may be covered.

The BOD_5 removal efficiency varies with the organic load imposed but usually fluctuates between 45 and 70% for a single-stage filter. Removal efficiencies of up to 90% can be achieved in two stages.

4.1.5 Roating biological contractors

Rotating biological contractors (RBC) units are another form of attached growth processes. In RBC units the biomass is attached to disks (up to 3.5 m in diameter) which rotate at 1 to 3 rpm while immersed up to 40% in the wastewater (Figure 4.13). The disks are made of corrugated, light plastic material.

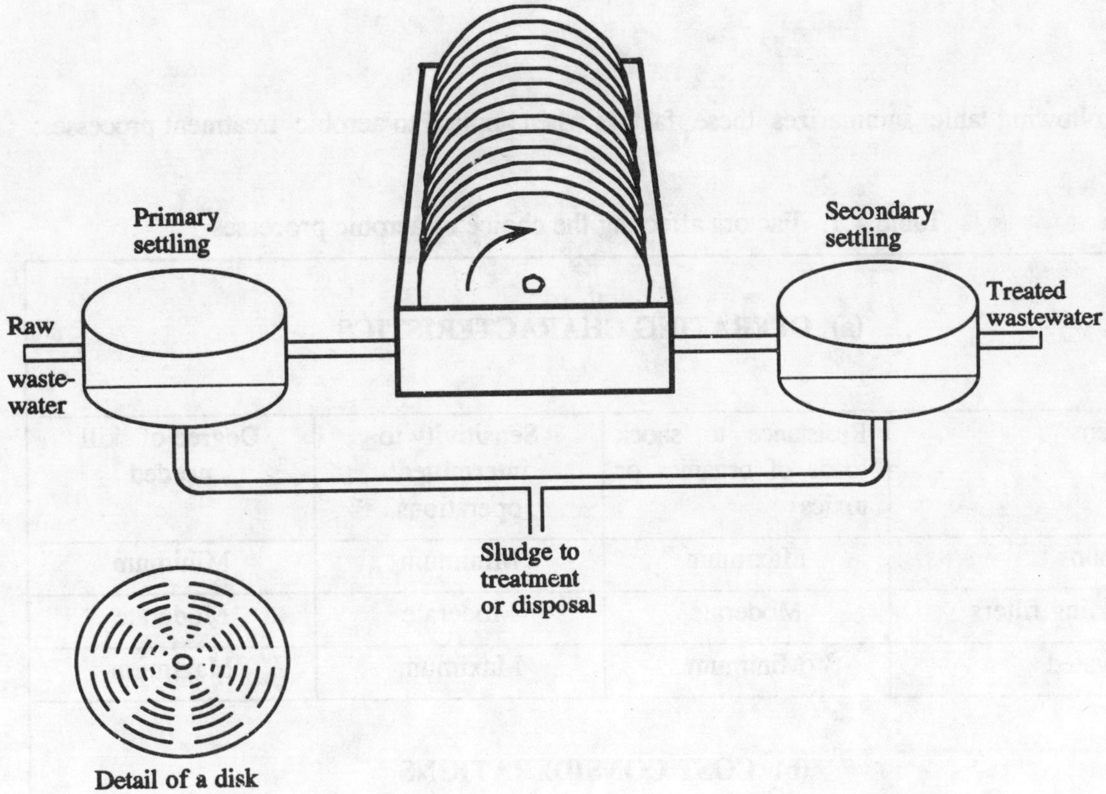

Figure 4.13. Diagram of rotating biological contractor (RBC) unit

When exposed to air the attached biomass absorbs air and when immersed the micro-organisms absorb the organic load. A biomass of 1-4 mm grows on the surface and its excess is teared off the disks by shearing forces and is separated from the liquid in the secondary settling tank. A small portion of the biomass remains suspended in the liquid within the basin and is also responsible in minor part for the organic load removal. Rotation speeds of more than 3 rpm are seldom used because this increases electric power consumption while the oxygen transfer does not increase sufficiently. The ratio of surface area of disks to liquid volume is typically 5 l/m^2. For high-strength wastewaters, more than one unit in series (staging) is used. The effect of lower temperatures is partially mitigated by the use of housing for the disk units. These systems are normally operated without recycling the liquid. The power consumption is in the order of 2 kW/1 000 m^3/day of capacity. They have been used to upgrade activated sludge existing plants, placing the disk units in the aeration basins (Antonie, 1978).

4.1.6 Selection of aerobic treatments

Several factors (apart from the economics) influence the choice of a particular aerobic treatment system. There is no universal solution and the decision of which system to use (or even if using an aerated system or not) depends on many aspects. Key factors are: the area available, which sometimes is the deciding aspect; the ability to operate intermittently is critical for several fishing industries which do not operate in a continuous fashion or work only seasonally; the skill needed for operation of a particular treatment cannot be neglected; and finally the costs (both operating and initial investment) are also sometimes decisive.

The following table summarizes these factors when applied to aerobic treatment processes:

Table 4.1. Factors affecting the choice of aerobic processes

(a) OPERATING CHARACTERISTICS

System	Resistance to shock loads of organics or toxics	Sensitivity to intermittent operations	Degree of skill needed
Lagoons	Maximum	Minimum	Minimum
Trickling filters	Moderate	Moderate	Moderate
Activated	Minimum	Maximum	Maximum

(b) COST CONSIDERATIONS

System	Land needed	Initial costs	Operating costs
Lagoons	Maximum	Minimum	Minimum
Trickling Filters	Moderate	Moderate	Moderate
Activated	Minimum	Maximum	Maximum

Adapted from Rich, 1980.

The considerations for the RBC systems are similar to those of trickling filters.

4.2 Anaerobic Treatment

The anaerobic treatment of wastewater proceeds with degradation of the organic load to gaseous products (mainly methane and carbon dioxide) which constitute most of the reaction products and biomass. Anaerobic treatment is the result of several reactions: the organic load present in the wastewater is first converted to soluble organic material which in turn is consumed by acid producing bacteria to give volatile fatty acids, plus carbon dioxide and hydrogen. The methane producing bacteria consume these to produce methane and carbon dioxide. This is summarized in Figure 4.14.

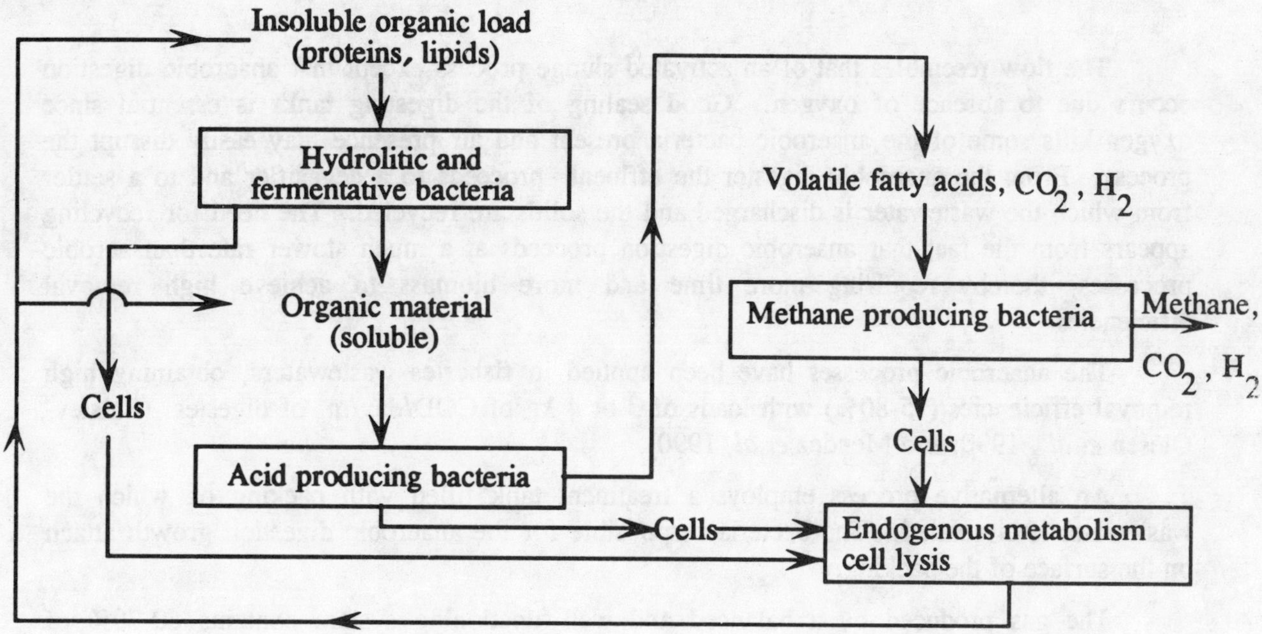

Figure 4.14. Scheme of reactions produced during anaerobic treatment

These processes are reported to be better applied to high-strength wastewaters (e.g., blood water or stickwater).

4.2.1 Digestion systems

A typical diagram is shown in Figure 4.15, which presents an anaerobic system.

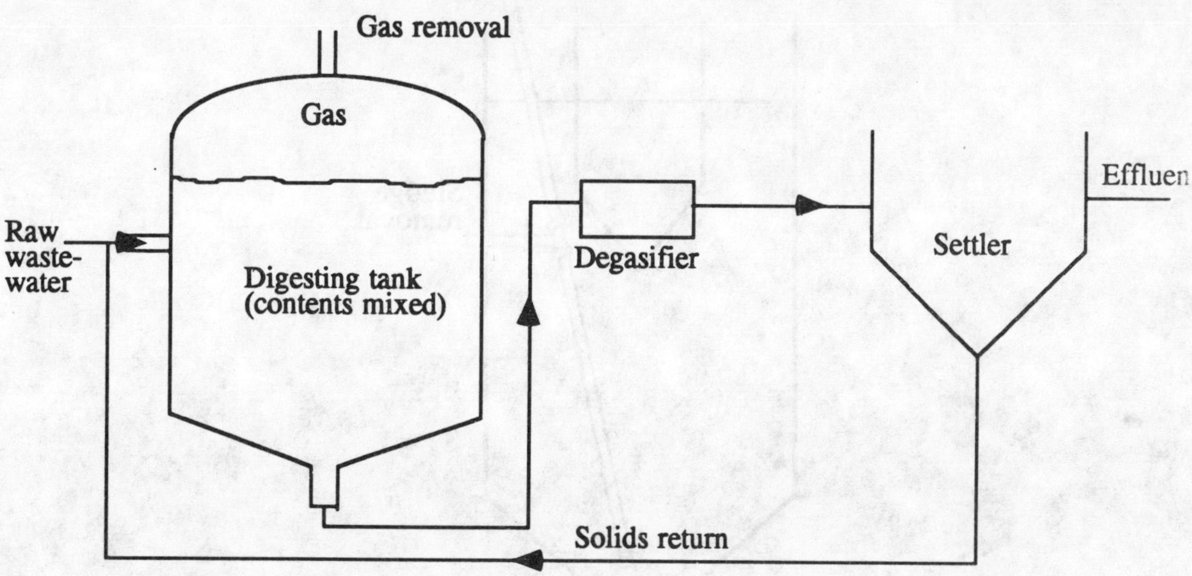

Figure 4.15. Diagram of an anaerobic digestion process

The flow resembles that of an activated sludge process except that anaerobic digestion occurs due to absence of oxygen. Good sealing of the digesting tanks is essential since oxygen kills some of the anaerobic bacteria present and air presence may easily disrupt the process. From the anaerobic digester the effluent proceeds to a degasifier and to a settler from which the wastewater is discharged and the solids are recycled. The need for recycling appears from the fact that anaerobic digestion proceeds at a much slower rate than aerobic processes, thereby requiring more time and more biomass to achieve high removal efficiencies.

The anaerobic processes have been applied in fisheries wastewaters, obtaining high remcval efficiencies (75-80%) with loads of 3 or 4 kg of COD/day/m^3 of digester (Balslev-Olesen *et al.*, 1990; and Mendez *et al*, 1990).

An alternative process employs a treatment tank filled with packing on which the wastewater is circulated. The bacteria responsible for the anaerobic digestion growth attach on the surface of the packing.

The gas produced by a balanced and well-functioning system contains 60-70% of methane, the rest being mostly carbon dioxide and minor amounts of nitrogen and hydrogen.

Anaerobic processes are also sensitive to temperature. This is why in some cases heating is provided to the digester to reach temperatures of 30°-35°C. In most cases this can be done in part with the methane gas originating from the digester.

4.2.2 Imhoff tanks

The Imhoff tank is a relatively simple system used originally instead of heated digesters. It is still used for plants of small capacities. It consists basically of a two-chamber rectangular tank, usually built partially underground (Figure 4.16).

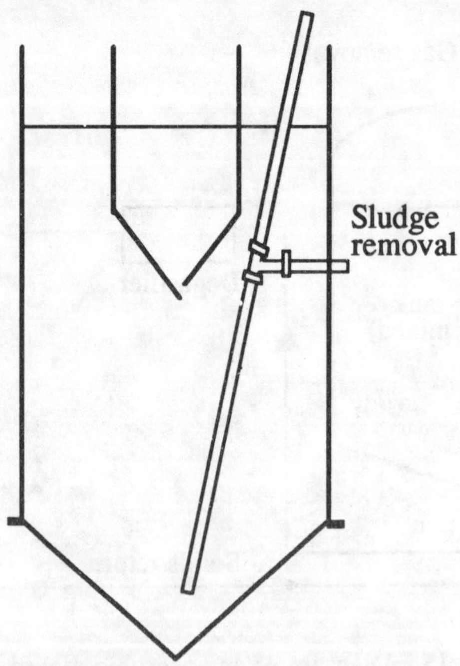

Figure 4.16: Section of an Imhoff tank

The wastewater enters in the upper compartment which acts as a settling basin while in the lower part the settled solids are stabilized anaerobically.

Short-circuiting of the wastewater can be prevented by using a baffle at the entrance together with more than one port for discharge.

The lower compartment is generally unheated. The stabilized sludge is generally removed from the bottom generally twice a year to give ample time for the sludge to stabilize, although the removal frequency is sometimes dictated by the convenience of sludge disposal. In some cases, these tanks are designed with inlets and outlets at both ends, and the flow of wastewater is reversed periodically so that the sludge in the bottom accumulates evenly. Although they are simple installations, they are not without inconveniences: foaming, odour and scum formation. These typically result when temperature falls below 15°C, when the bacteria that produce volatile acids (see Figure 4.14) predominate and methane production is reduced causing a process imbalance. This is why in some cases immersion heaters are used during cold weather. The scum forms because the gases that originate during the anaerobic digestion are entrapped by the solids and have no time to escape from the solids causing them to float. This is usually overcome by increasing the depth in the lower (digestion) chamber. At lower depths, bubbles form at a higher pressure, expand more when rising and are more likely to escape from the solids. The odour problems are minimal when the two stages of the process (acid formation and gas formation) are balanced.

5. PHYSICOCHEMICAL TREATMENTS

5.1 Coagulation - Flocculation

In coagulation operations, a chemical substance is added to an organic colloidal suspension to cause its destabilization by the reduction of forces that keep them apart. It involves the reduction of surface charges responsible for particle repulsions. This reduction in charge causes flocculation (agglomeration). Particles of larger size are then settled and clarified effluent is obtained. A diagram of a coagulation-flocculation and settling of a wastewater is shown in Figure 5.1.

In fisheries wastewaters, the colloids present are of organic nature and are stabilized by layers of ions that result in particles with the same surface charge, thereby increasing their mutual repulsion and stabilization of the colloidal suspension. This kind of wastewater may contain appreciable amounts of proteins and micro-organisms which become charged due to the ionization of carboxyl and amino groups or their constituent amino acids. The grease and oil particles, which are normally neutral, become charged due to preferential absorption of anions (mainly hydroxyl ions). The parameter that characterizes the stability of a particle is called the Z potential and is a measure of the potential that would be required for destabilization of the particles. However, the usefulness of the Z potential is often questioned since it varies with the composition of the solution and is not repeatable

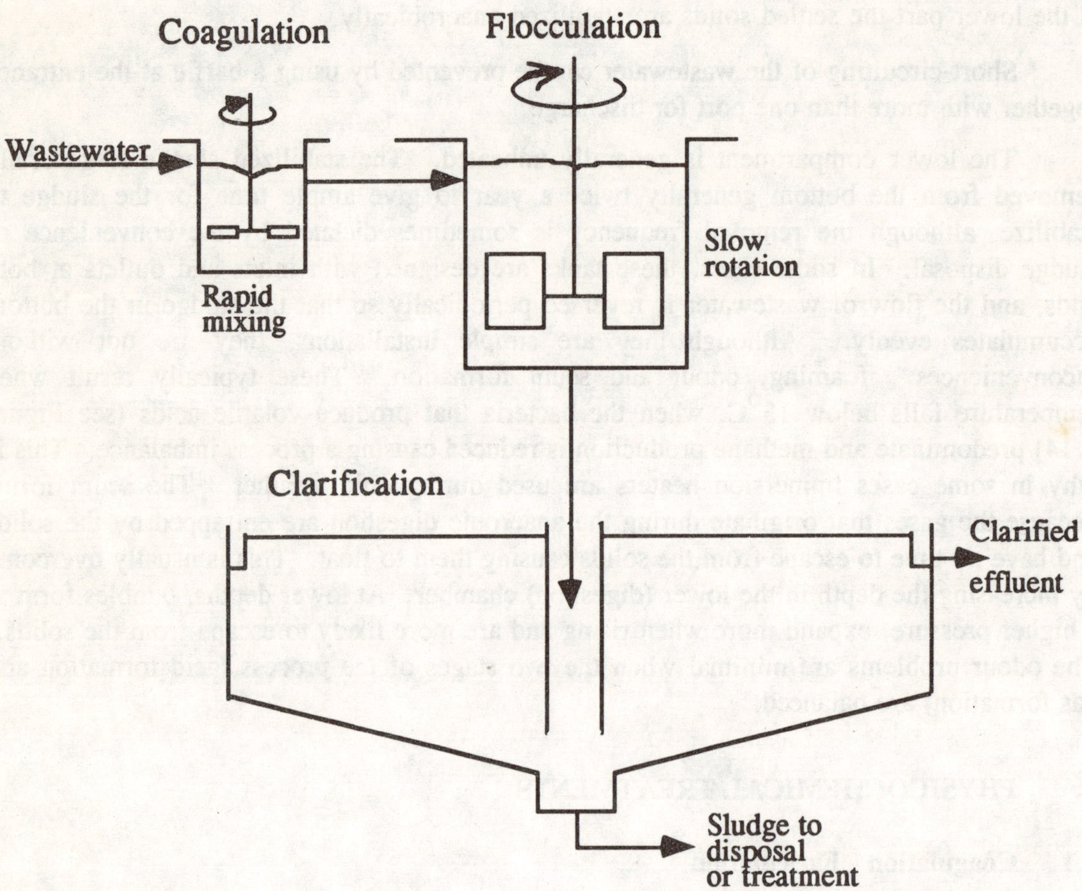

Figure 5.1. Simplified diagram of a chemical coagulation process

The coagulation processes generally employ several steps. First, coagulant is added to the effluent, and mixing proceeds rapidly and with high intensity. The objective is to obtain intimate mixing of the coagulant with the wastewater, increasing the effectiveness of destabilization of particles and initiating coagulation. A second stage follows in which flocculation occurs for periods of up to 30 minutes. In the latter case, the suspension is stirred slowly to increase the possibility of contact between coagulating particles and to facilitate the development of large flocs. These flocs are then transferred to a clarification basin in which they settle and are removed from the bottom while the clarified effluent overflows.

Several substances may be used as coagulants. Given the proteinaceous nature of several wastewaters, pH can be adjusted by adding acid or alkali. The first is more common and causes coagulation of the proteins by denaturing them, changing their structural conformation due to the change in charge distribution on their surface. Thermal denaturation of proteins can also be used but due to its high energy demand it is only advisable if excess steam is available. In fact, the "cooking" of the blood-water in fishmeal plants is basically a thermal coagulation process.

Polyelectrolites are also commonly used as coagulants. They act as coagulants by lowering the charge of the wastewater particles. For this case, cationic polyelectrolites are preferred since wastewater particles generally have a negative charge. Some polyelectrolytes act as "bridges" between the already-formed particles. In the flocculation process, bridged particles interact with other bridged particles and this results in the increase of floc size. For this purpose, anionic or neutral polyelectrolites are used.

Since the recovered sludges from coagulation-flocculation processes may sometimes be added in the formulation of animal feeds, it is advisable to check that the coagulant or flocculant used are not toxic.

In fisheries wastewaters there are several reports of the use (at both pilot plant and at working scale) of inorganic coagulants such as aluminum sulphate and ferric chloride or ferric sulphate as well as organic coagulants (Johnson, 1984; Nishide 1976 and 1977; Ziminska, 1985).

There are reports (Hood and Zall, 1980) that fish scales can be used effectively as an organic wastewater coagulant. For this purpose, they are dried and ground before being added as coagulant in powder form. Another by-product of marine origin that is used as coagulant is chitosan, a natural polymer derived from chitin, a main constituent of the exoskeletons of crustacea.

No single set of operating conditions will satisfy all kinds of fisheries wastewaters and, therefore, the operational policy, pH, and coagulant dosage must be determined for each individual case. This is generally done in the laboratory using beakers and stirrers, trying to simulate the operation at full scale: first, the coagulant is added and rapidly mixed, then stirring proceeds slowly to allow flocculation. Since several variables are studied at a time, it is common to use multiple places (usually 4 or 6) stirrers. These allow constant stirring conditions to be maintained and therefore ascertain the effect of coagulant dosage and pH which is varied in the beakers. The device is commonly known as the "Jar Test apparatus" (see Figure 5.2).

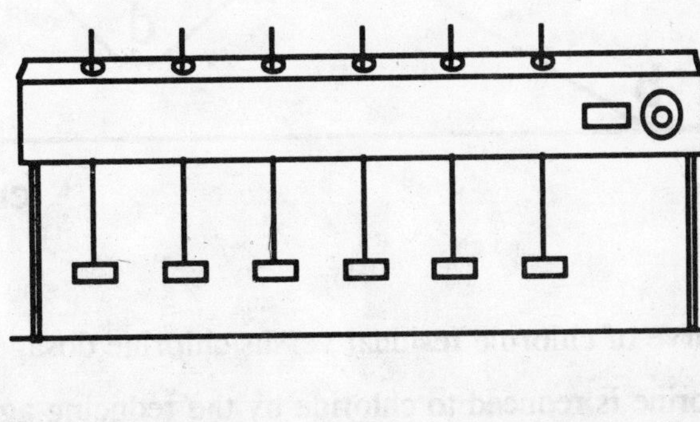

Figure 5.2. Typical Jar Test apparatus, normally provided with a single speed control and indicator, and each stirrer can be either disengaged or placed at different heights

5.2 Disinfection

5.2.1. Chlorination

Chlorination is a process commonly used in both industrial and domestic wastewaters. Although there may be other reasons for its use in other kinds of wastewaters (e.g., cyanide oxidation), the reason for its use in fisheries effluents is the disinfection by destroying bacteria or algae or inhibiting their growth. Usually the effluents are chlorinated just before their final discharge to the receiving waterbodies.

For this process either chlorine gas or hypochlorite solutions may be used, the latter being easier to handle.

In water solutions, chlorine forms hypochlorous acid which in turn forms hypochlorite:

$$Cl_2 + H_2O \leftrightarrow HOCl + H^+ + Cl^-$$
$$HOCl \leftrightarrow H^+ + OCl^-$$

A problem that may occur during chlorination of fisheries effluents is the formation of chloramines. These wastewaters may contain appreciable amounts of ammonia or volatile amines, which cause an increased demand of chlorine to achieve a desired degree of disinfection because they react with chlorine to give chloramines. The relative proportions of these products depends on the pH and concentration of ammonia and the organic amines present.

The degree of disinfection is attributed to the residual chlorine present in water. A typical plot of the residual chlorine versus the chlorine dosage is shown in Figure 5.3:

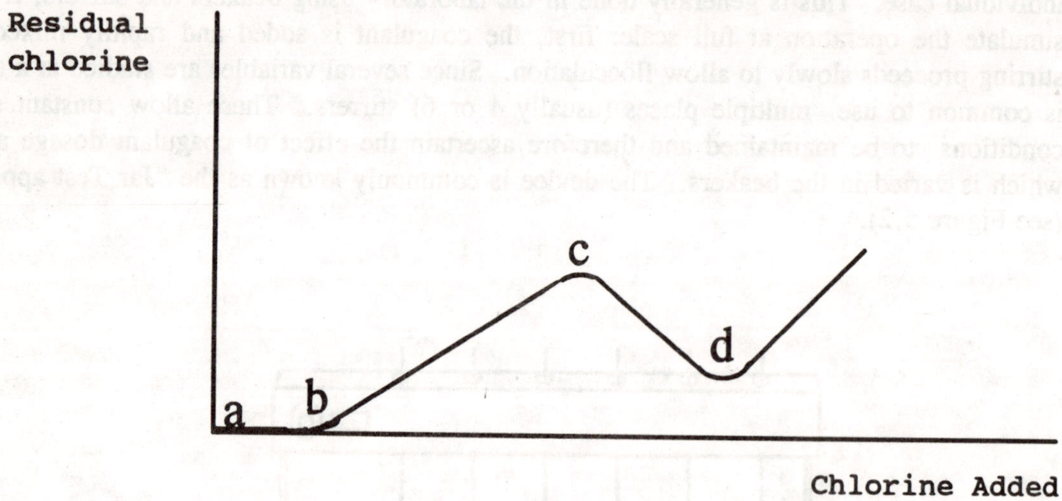

Figure 5.3. Curve of chlorine residual versus chlorine dosage

An initial amount of chlorine is reduced to chloride by the reducing agents present, and therefore the residual chlorine is negligible (segment a-b). Further addition of chlorine may result in the formation of chloramines. These appear as residual chlorine but in the form of combined chlorine residual (segment b-c). Once all the ammonia and organic amines have reacted with the added chlorine, additional amounts of chlorine result in the disappearance of the chloramines by oxidation, with a decrease in the chlorine residual as a consequence (segment c-d). Once this oxidation is complete, further addition of chlorine results in the appearance of free available chlorine. The point d on the curve is also known as "breakpoint". The goal in obtaining some free chlorine residual is to ensure disinfection.

Chlorination units are generally simple, consisting of a chlorination vessel in which the wastewater and the chlorine are brought into contact. Sufficient mixing must be provided, and it is recommended that the residence time be not less than 30 seconds. Prior to final discharge, sufficient time (about 15 minutes) must be allowed for the chlorine to react. This may be done in the ducts which carry the wastewater to the discharge point, provided the residence time exceeds 15 minutes. Otherwise, a contact basin must be constructed. A typical configuration is shown in Figure 5.4.

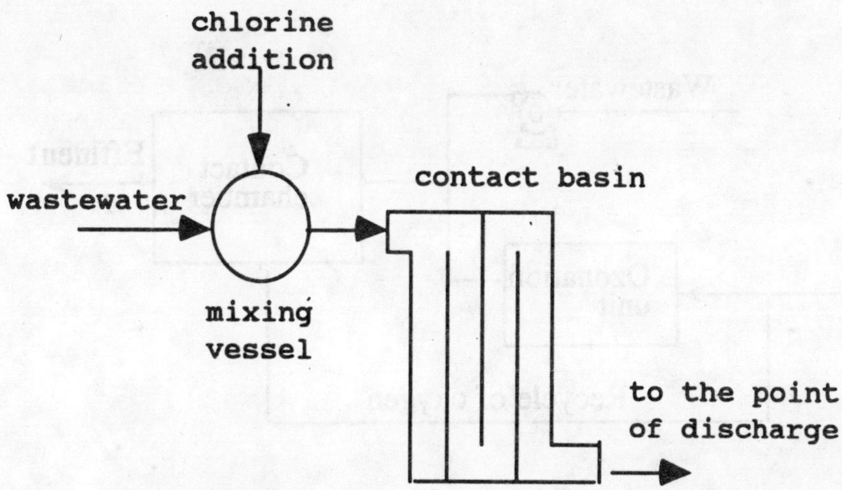

Figure 5.4. Schematics of a chlorination system

The channels in this contact basin are usually narrow in order to increase the water velocity. This reduces accumulation of solids by settling. However, the space between the channels should allow for easy cleaning.

The levels of available chlorine after the breakpoint should comply with the local regulations that usually vary between 0.2 mg/l and 1 mg/l. This value strongly depends on where the wastewater is to be discharged because residual chlorine in wastewater effluents was identified in some cases as the main toxicant suppressing the diversity, size, and quantity of fish in receiving streams (Paller *et al.*, 1983).

Although 15 minutes is very common, retention times of up to 30 minutes are also used. The chlorine dosage needed to achieve the residuals required will vary with the wastewater considered: 2-8 mg/l is common for an effluent from activated sludge plant, and can be as high as 40 mg/l in the case of septic wastewater (Eckenfelder, 1980, Metcalf and Eddy Inc., 1979).

5.2.2 Ozonation

Since ozone is also a strong oxidizing agent, it has been used for disinfection due to its bactericidal properties and its potential for removal of viruses. Ozone (O_3) is produced when a high voltage is discharged across a narrow gap in the presence of air or oxygen.

Ozonation systems generate O_3 *in situ*. Figure 5.5 shows a diagram of an ozonation system.

Once ozone has been added and reacted, it reverts to oxygen increasing somewhat the dissolved oxygen level of the effluent to be discharged, which is beneficial to the eceiving water stream. Contact tanks are usually closed to recirculate the oxygen-enriched air to the ozonation unit. Advantages over chlorination are that it does not produce dissolved solids and is affected neither by ammonia compounds present nor by the pH value of the effluent. It is also used to oxidize ammonia and nitrites present in fish culture facilities (Monroe and Key, 1980).

Although much less used than chlorination in fisheries wastewaters, ozonation systems have been installed especially in discharges to sensitive waterbodies (Stover and Jarnis, 1979; Rosenthal and Kruner, 1985).

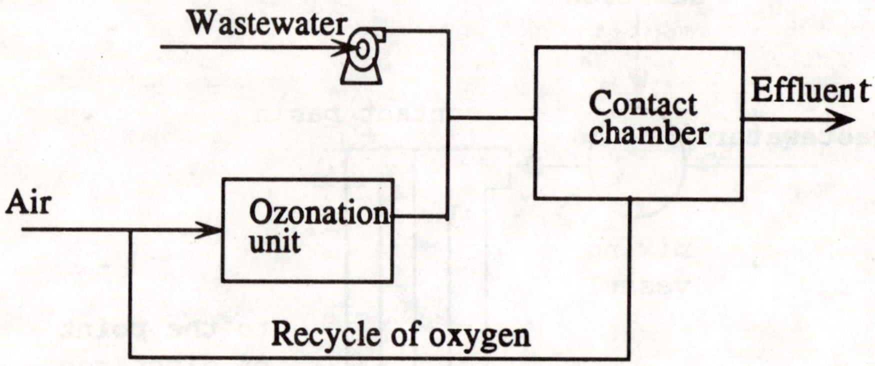

Figure 5.5. Diagram of an ozonation system

6. SLUDGE TREATMENT AND DISPOSAL

The sludge generated in the primary and secondary treatments is a suspension of solids in concentrations ranging from about 1 to 10%, depending on the treatment from where it originates. It contains almost all of the organic load removed from the treated wastewater. Although it is in a different form, it is organic matter that may eventually decompose, be offensive and be a contaminant if not treated properly.

In this section, only the most simple treatments will be discussed. The description of more elaborate processes such as chemical conditioning, heat treatment, machine-drying and thermal reduction or incineration can be found elsewhere (Metcalf and Eddy Inc., 1979; Eckenfelder, 1980). Some of the treatments described here may be applicable only by fisheries which have large processing capacities (e.g., digestion), while others (hauling) are more appropriate for small-scale, intermittent operating plants.

6.1 Sludge Transportation to Treatment Facilities

The transportation of the sludge generated in plants to sludge treatment facilities is conceptually simple. Since the sludges from fish processing plants are highly biodegradable they will probably be accepted without problems. For small fish processors this solution may only be adopted due to the lack of space for sludge treatment. Other advantages are that no investment is needed for the treatment units and the only operating costs are those of transportation and the fees charged by the remote sludge treatment facility.

6.2 Sludge Digestion

Sludges are residues with relatively high carbon content and are good candidates for anaerobic digestion as they are wastes with high organic content. The digesters are very similar to those of anaerobic treatment of wastewaters described previously in section 4. Here, the cells produced by the anaerobic treatment undergo hydrolysis and enter the process depicted in Figure 4.14. For sludge digestion, two processes are the most common: the conventional digestion and the high rate digestion, as shown in Figure 6.1.

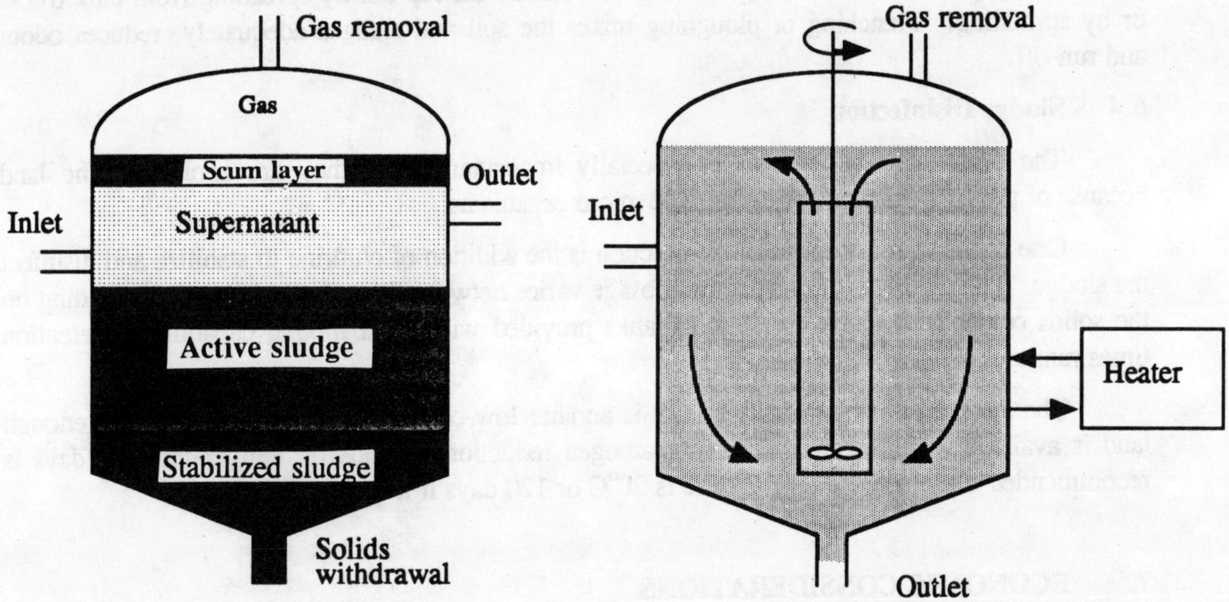

Figure 6.1. Conventional (left) and high rate (right) sludge digestion processes

In the conventional digestion process, sludge digestion and sludge thickening are carried out in the same vessel. The sludge enters the digester in the zone where active digestion is taking place. Due to the conversion of biomass to gases, the sludge stabilizes and settles to the bottom. The gas formed accumulates in the upper part of the digester and, while it rises, carries small particles of oil/grease which form a scum layer on top. Here, the goal is to stabilize the sludge, converting most of it into gas. Typical retention times range from 30 to 60 days and the contents are normally heated to near 35°C in order to accelerate the anaerobic reactions which otherwise would require much more time.

In the high rate sludge digestion, the sludge is mixed by recirculating the gas formed in a draft tube or by pumping. Typical power consumption for mixing ranges from 0.006 to 0.01 kW/m^3. Retention times in the digester are shorter (10-20 days) than in the conventional process and higher organic loads can be used. Due to continuous mixing, this process requires a settling tank after the digestion stage.

6.3 Land Disposal of Sludges

This is one of the lower cost alternatives for final sludge disposal. It is practised for two reasons: the need to dispose of the sludge (itself also a residue) and the opportunity to use the contents (organic matter, nitrogen and phosphorus) as fertilizer. Sludge land application has proven to be a viable substitute for commercial fertilizer, provided pathogenic control is applied.

Site selection is an important factor; it should be relatively far from waterbodies and the soil should be moderately permeable and well-drained, with a maximum slope of 5-8% in order to minimize erosion problems during application. Neutral to alkaline soils are preferred to reduce heavy metal mobility, the concentration of which should also be periodically checked.

The application of the sludges on the soil can be carried out by spreading from tank trucks or by sprinkling. Trenching or ploughing mixes the soil and sludges adequately, reduces odour and run-off.

6.4 Sludge Disinfection

The disinfection of sludges is especially important before they are applied to the land because of possible existence of pathogenic micro-organisms.

One of the simplest forms of disinfection is the addition of chlorine to stabilize and disinfect the sludge. The chlorine for disinfection dosage varies between 2 mg/l and 15 mg/l, depending on the solids content. It is usually done in tanks provided with good mixing capability. Retention times range from 15 to 45 minutes.

The storage of sludge for long-term is another low-cost practice, applicable only if enough land is available. The storage time for pathogen reduction depends on temperature: 60 days is recommended if the average temperature is 20°C or 120 days if it is only 4°C.

7. ECONOMIC CONSIDERATIONS

After the preliminary selection of which process to apply, economic considerations are among the most important parameters that influence the final decision as to which process should be chosen for wastewater treatment.

To develop cost estimates, the data from the wastewater characterization should be available together with the design parameters for alternative processes and the cost correlations for these. The costs related to these alternative processes and information on the quality of effluent should also be obtained prior to the elaboration of cost estimates before discharge in compliance with local regulations.

7.1 Cost of Unitary Processes

Table 7.1 lists the construction costs for the most common unitary processes of wastewater treatment. These were developed for municipal sewage treatment and may not be totally applicable for very small wastewater treatment plants but are useful for preliminary estimation and comparison among alternatives.

Table 7.1. Construction costs for selected unitary operations of wastewater treatment adapted from EPA (1978)

Liquid stream	Correlation
Preliminary treatment	$C = 5.79 \times 10^4 \times Q^{1.17}$
Flow equalization	$C = 1.09 \times 10^5 \times Q^{0.49}$
Primary Sedimentation	$C = 1.09 \times 10^5 \times Q^{1.04}$
Activated Sludge	$C = 2.27 \times 10^5 \times Q^{0.87}$
Rotating Biological Contactor	$C = 3.19 \times 10^5 \times Q^{0.92}$
Chemical Addition	$C = 2.36 \times 10^4 \times Q^{1.68}$
Stabilization Pond	$C = 9.05 \times 10^5 \times Q^{1.27}$
Aerated Lagoon	$C = 3.35 \times 10^5 \times Q^{1.13}$
Chlorination	$C = 5.27 \times 10^4 \times Q^{0.97}$
Solids stream	**Correlation**
Sludge Handling	$C = 4.26 \times 10^4 \times Q^{1.36}$
Aerobic Digestion	$C = 1.47 \times 10^5 \times Q^{1.14}$
Anaerobic Digestion	$C = 1.12 \times 10^5 \times Q^{1.12}$
Incineration	$C = 8.77 \times 10^4 \times Q^{1.33}$

7.2 Cost of Operation and Maintenance

The main factors that influence the costs of operation and maintenance are: energy costs (power demand); labour costs (which should include the personnel for operation, maintenance and administrative services); materials costs; cost of chemicals (sometimes included in the previous item); and cost of transportation of sludges for final disposal and discharge of treated wastewater.

The relative importance of these items varies significantly depending on the location, the quality of the effluent discharged and on the specific characteristics of the wastewater being treated.

7.3 Capital Costs

These comprise mainly the unit construction costs, the land costs, the cost of the treatment units, and the cost of engineering, administration and contingencies.

The location should be carefully evaluated in each case because it affects the capital costs more than the operating costs.

The cost of equipment may also be a significant portion of the capital costs in more automated installations.

A recent compilation of capital costs for several processes of physical treatment has been presented by Wright and Woods (1993), where capital cost correlations are given for oil-water separators, equalization basins, primary clarifiers, secondary clarifiers, a reverse osmosis unit, ultrafiltration units, gravity filters and microscreens.

7.4 Estimation of Total Costs

For small wastewater treatment plants, an initial estimate of the total cost can be obtained from the cost of a similar plant with a different capacity, a relationship derived from costs relationships in chemical industries. The cost of plants of different sizes is related to the ratio of their capacity raised to the 0.6 power:

$$\text{Capital cost 2} = \text{Capital cost 1} \times \left(\frac{\text{Capacity 2}}{\text{Capacity 1}}\right)^{0.6}$$

where Capital cost 1, 2 = capital costs of plants 1 and 2
Capacity 1, 2 = capacity of plants 1 and 2

The operation and maintenance costs can be estimated by a similar formula:

$$\text{O.M. 2} = \text{O.M. 1} \times \left(\frac{\text{Capacity 2}}{\text{Capacity 1}}\right)^{0.85}$$

where: O.M. 1, 2 1, 2 = operation and maintenance costs of plants 1 and 2
Capacity 1, 2 = capacity of plants 1 and 2

An alternative procedure for the development of cost models for wastewater treatment systems includes the preparation of kinetic models for the possible treatment alternatives, in terms of area and flow rates at various treatment efficiencies, followed by the computation of mechanical and electrical equipment, as well as the operation and maintenance costs as a function of the flow rates (Ulatam, 1991).

The models so developed can be used to select the most appropriate treatment process.

8. REFERENCES[1]

AGOS (Administracion General de Obras Sanitarias), 1990. Province of Buenos Aires, Argentina, Rule No. 287, April 9th, 1990.

APHSM (American Public Health Association), 1989. Standard Methods for the Examination of Water and Wastewater 17th.. Ed.,Washington DC, USA.

Anonymous, 1986. Environmental assessment and management of the fish processing industry, Sectoral studies series Nr 28, UNIDO, Vienna, Austria.

Antonie, R.L.; 1978. Design criteria for applications of the rotating biological contactor to domestic and industrial wastewater treatment, International Environment Colloquium, May 16-17, Liege, Belgium.

API (American Petroleum Institute), 1959. Manual for disposal of refinery wastes; API, New York, U.S.A.

Bauman, F.J.; 1974. Dichromat Reflux Chemical Oxygen Demand. A Proposed Method for Chlorine Correction in Highly Saline Wastes. Analytical Chemistry 46:1336-1338.

Balslev-Olesen, P., Lynggaard, A. and Nickelsen, C.; 1990. Pilot-Scale experiments on Anaerobic Treatment of Waste-water from a Fish Processing Plant. Water Science and Technology 22:463-474.

Civit, E.M; Parin, M.A. and Lupin, H.M.; 1982. Recovery of Protein and Oil from Fishery Blood-water Waste. Effect of pH and Temperature. Water Research, 16:809-814.

Cuadros, M. and Gonzalez, J.F.; 1991. Study of the Environmental Impact of the Fishing Industry in the Ferrol Bay at Chimbote, Perú. Pub. Nr 87.2145.8-03.100. Gesselschaft für Technische Zusammenarbeit, Hamburg, Germany.

Eckenfelder, W. W. and Ford, D.L.; 1969. Laboratory and design procedures for waste treatment process design; Texas University Press, Austin, Texas; U.S.A.

Eckenfelder, W. W; 1980. Principles of Water Quality Management; CBI Publishing Co.; Boston, U.S.A.

EPA (Environmental Protection Agency), 1978. Construction costs for municipal wastewater treatment plants: 1973-1977. Technical Report MCD-37, U.S. E.P.A. Washington, D.C., U.S.A.

---

[1] These references are presented as submitted by the author.

Ertz, D.B; 1980. Dissolved Air Flotation Treatment of Seafood Processing Wastes. An Assessment, in Proc. 8th. Natl.Symp. on Food Processing Wastes, Cincinnati, Ohio, U.S.A.

Gonzalez, J.F.; 1985. Fellowship Report to CONICET (Consejo Nacional de Investigaciones Cientificas y Tecnicas), Buenos Aires, Argentina.

Gonzalez, J.F; 1986. Semi-Micro Determination of COD on Fish Filleting Wastewater; Environmental Technology Letters; 7:269-272.

Gonzalez, J.F.; Civit, E.M. and Lúpin, H.M.; 1983. Composition of Fish Filleting Wastewater; Water South Africa; 9:49-56.

Hansen, S.P. and Culp, G.L.; 1967. Applying Shallow Depth Sedimentation Theory; Journal of the American Water Works Association 59:1134-1148.

Herborg, L; 1974. Wastewater from the Fishing Industry; FAO (Food and Agriculture Organization), Rome, Italy.

Hood, L.F. and Zall, R.R.; 1980. Recovery, Utilization and Treatment of Seafood Processing Wastes, in Advances in Fish Science and Technology, J.J. Conell, Ed. Fishing News Books, Ltd. Surrey, England.

Ilet, K.J.; 1980. Dissolved Air Flotation and Hydrocyclones for Wastewater Treatment and By-product Recovery in the Food Process Industries; Water Services 84:26-27.

Jirka, A.M. and Carter, M.J.; 1975. Micro Semi Automated Analysis of Surface and Waste Waters for Chemical Oxygen Demand; Analytical Chemistry; 47:1397-1402.

Johnson, R.A and Gallager, S.M.; 1984. Use of Coagulants to treat Seafood Processing Wastewaters; Journal Water Pollution Control Federation 56:970-976

Mendez, R.; Omil, F.; Soto, M. and Lema, J.M.; 1992. Pilot Plant studies on the Anaerobic Treatment of Different Wastewaters from a Fish Canning Factory; Water Science and Technology 25:37-44.

Metcalf and Eddy Inc; 1979. Waste-water Engineering: Treatment, disposal, reuse. McGraw-Hill Book Co, New York, U.S.A.

Middlebrook, E.J., 1979. Industrial Water Pollution Control; J. Wiley & Sons, New York, U.S.A.

Monroe, D.W. and Key, W. P.; 1980. The Feasibility of Ozone for Purification of Hatchery Waters;Ozone Science and Engineering;2: 203-224.

Nemerow, N. L.; 1971. Liquid Wastes of Industry; Addison-Wesley Publishing Co., Reading, Mass; U.S.A.

Nishide, E.; 1976. Coagulation of Fishery wastewater with Inorganic Coagulants; Bulletin of the College of Agriculture and Veterinary Medicine, Nihon University, Japan 33:468-475.

Nishide, E.; 1977. Coagulation of Fishery wastewater with Inorganic Coagulants; Bulletin of the College of Agriculture and Veterinary Medicine, Nihon University, Japan 34:291-294.

Paller, M. H. and Lewis, W. M.; R.C. Heidinger and Wawronowicz, J.L.; 1983. Effects of Ammonia and Chlorine on Fish in Streams Receiving Secondary Discharges.; Journal Water Pollution Control Federation; 55:1087-1097.

Parin, M.A., Civit, E. M. and Lupín, H.M.; 1979. Characterization of Blood-water Effluent of Fish Meal Factories. Revista Latinoamericana de Ingenieria Quimica y Quimica Aplicada; 9:155-164.

Ramalho, R.S.; 1977. Introduction to Wastewater Treatment Processes; Academic Press, Inc.; New York, U.S.A.

Rich, L.G.; 1980. Low Maintenance, Mechanically Simple Wastewater Treatment Systems, McGraw-Hill Book Co., New York; U.S.A.

Rosenthal, H. and Kruner, G.; 1985. Efficiency of an Improved Ozonation Unit Applied to Fish Culture Situations; Ozone: Science and Engineering 7:179-190.

Stanier, R. Y.; Adelberg, E. A. and Ingraham, J.; 1985. The Microbial World 4th. Ed.. Prentice Hall, Englewood Cliffs, New Jersey, U.S.A.

Stover, E. L. and Jover, R.N.; 1980. High Level Ozone Disinfection of Wastewater for Shell Fish Discharges; Ozone Science and Engineering; 1:335-346.

Sorensen, P.E.; 1974. Fiskeriindustri-Vandproblemer; Vandkvalitetinstituttet, Copenhagen, Denmark.

Uluatam, S. S.; 1991. Cost models for small wastewater treatment plants. International Journal of Environmental Studies; 37:171-181.

Wright, D.G. and Woods, D.R.; 1993. Evaluation of capital cost data. Part 7. Liquid waste disposal with emphasis on physical treatment; Canadian Journal of Chemical Engineering 71:575-590.

Ziminska, H.; 1985. Protein Recovery from Fish Wastewaters; Proceedings of the Fifth International Symposium on Agricultural. Wastes; p379;American Society of Agricultural Engineering; St. Joseph, MI, U.S.A.

9. INDEX